TEST BANK

LuAnne Clark
Lansing Community College

to accompany

Introduction to the Human Body

The Essentials of Anatomy and Physiology

Sixth Edition

Gerard J. Tortora
Bergen Community College

Sandra Reynolds Grabowski
Purdue University

WILEY

JOHN WILEY & SONS, INC.

COVER PHOTO: Paul Taylor Dance Company in "Field of Grass." Photo ©1998 Lois Greenfield.
Reproduced with permission.

To order books or for customer service call 1-800-CALL-WILEY (225-5945).

ISBN 0-471-42071-9

Printed in the United States of America.

10 9 8 7 6 5 4 3 2 1

Printed and bound by Victor Graphics, Inc.

Table of Contents

Chapter 1 Organization of the Human Body

Multiple-Choice

Choose the one alternative that best completes the statement or answers the question.

1. The science dealing with the functions of the body parts is called
   A) physiology.
   B) cytology.
   C) anatomy.
   D) histology.
   E) biology.
   Answer: A

2. The level of organization when different tissues join together is called the
   A) chemical level.
   B) cellular level.
   C) tissue level.
   D) organ level.
   E) system level.
   Answer: D

3. Group of related organs that have a common function is called a(n)
   A) organ.
   B) system.
   C) tissue.
   D) group.
   E) organism.
   Answer: B

4. The process by which unspecialized cells become specialized cells is called
   A) anabolism.
   B) catabolism.
   C) metabolism.
   D) differentiation.
   E) homeostasis.
   Answer: D

5. The sum of all chemical processes in our body is
   A) anabolism.
   B) catabolism.
   C) metabolism.
   D) differentiation.
   E) homeostasis.
   Answer: C

6. All of the following are examples of organs EXCEPT
   A) stomach.
   B) heart.
   C) muscle.
   D) brain.
   E) gallbladder.
   Answer: C

7. The condition in which the body's internal environment stays within physiological limits is
   A) responsiveness.
   B) homeostasis.
   C) differentiation.
   D) growth.
   E) All of the above are correct.
   Answer: B

8. In a negative feedback system, the response of the effector
   A) enhances the original stimulus.
   B) eliminates the original stimulus.
   C) reverses the original stimulus.
   D) does not change the original stimulus.
   E) does not affect the original stimulus.
   Answer: C

9. Reproduction refers to
   A) increase in the size of the cells.
   B) the formation of new cells.
   C) production of a new individual.
   D) Both B and C are correct.
   E) Both A and B are correct.
   Answer: D

10. A structure which sends inputs to the control center is a(n)
   A) effector.
   B) receptor.
   C) affector.
   D) stimulus.
   E) output structure.
   Answer: B

11. A baroreceptor measures
   A) touch.
   B) blood pressure.
   C) temperature.
   D) alveolar stretch.
   E) pain.
   Answer: B

12. Which of the following is a symptom of disease rather than a sign?
   A) nausea
   B) bleeding
   C) vomiting
   D) fever
   E) rash
   Answer: A

13. The science that deals with the drug treatment of disease is called
   A) physiology.
   B) anatomy.
   C) epidemiology.
   D) pharmacology.
   E) pharmacy.
   Answer: D

14. In the anatomical position the subject
    A) is lying down.
    B) has arms placed above the head.
    C) has arms folded on the chest.
    D) is standing upright facing the observer with the palms backwards.
    E) is standing upright facing the observer with the palms forward.
    Answer: E

15. The plane that divides the body into unequal left and right portions is the
    A) parasaggital plane.
    B) midsaggital plane.
    C) frontal plane.
    D) transverse plane.
    E) oblique plane.
    Answer: A

16. A gluteal injection refers to an injection into the
    A) buttock.
    B) lower limb.
    C) ankle.
    D) upper limb.
    E) trunk.
    Answer: A

17. The sternum (breast bone) is ___________ to the heart.
    A) posterior
    B) anterior
    C) inferior
    D) lateral
    E) distal
    Answer: B

18. The plane that divides the body into a superior and inferior portion is the
    A) parasaggital plane.
    B) midsaggital plane.
    C) transverse plane.
    D) oblique plane.
    E) frontal plane.
    Answer: C

19. Which of the following is NOT considered a basic tissue type?
    A) connective tissue
    B) epithelial tissue
    C) cartilage tissue
    D) nervous tissue
    E) muscle tissue
    Answer: C

20. The organ system which regulates the body's activities through hormones is the
    A) digestive system.
    B) endocrine system.
    C) nervous system.
    D) cardiovascular system.
    E) integumentary system.
    Answer: B

21. The ___________ is the region between the lungs from the breastbone to the vertebra.
A) vertebral canal
B) pericardium
C) mediastinum
D) pleural cavity
E) manubrium
Answer: C

22. The organ system that transports nutrients from the gastrointestinal tract to the cardiovascular system is the
A) digestive system.
B) endocrine system.
C) lymphatic system.
D) urinary system.
E) respiratory system.
Answer: C

23. The anatomical term which best describes a structure toward the head is
A) superficial.
B) deep.
C) inferior.
D) superior.
E) anterior.
Answer: D

24. The ring finger is _______________to the little and middle fingers.
A) lateral
B) intermediate
C) deep
D) distal
E) proximal
Answer: B

25. The anatomical term for navel is
A) crural.
B) inguinal.
C) umbilicus.
D) femoral.
E) coxal.
Answer: C

26. The best anatomical term to describe the back region of the body would be
A) ventral.
B) dorsal.
C) gluteal.
D) vertebral.
E) popliteal.
Answer: B

27. The brain and the spinal cord are located in the
A) ventral cavity.
B) dorsal cavity.
C) abdominal cavity.
D) thoracic cavity.
E) vertebral cavity.
Answer: B

28. All of the following belong to the ventral body cavity EXCEPT the
   A) thoracic cavity.
   B) abdominal cavity.
   C) cranial cavity.
   D) pleural cavity.
   E) pericardial cavity.
   Answer: C

29. The body demonstrates movement as a characteristic of life through the process of
   A) digestion.
   B) eating.
   C) breathing.
   D) excretion.
   E.) walking.
   Answer: E

30. Which is the most correct level of organization from the smallest to the largest?
   A) Atom – Element – Organ – Organelle – Cell – Tissue
   B) Atom – Molecule – Organ – Tissue – Cell
   C) Molecule – Atom – Organelle – Cell – Tissue
   D) Atom – Molecule – Organelle – Cell – Organ
   E) Molecule – Atom – Organelle – Cell – Organ
   Answer :D

31. The human body's ability to respond to changes in either its internal or external environment is called
   A) reproduction.
   B) responsiveness
   C) adaptation.
   D) metabolism.
   E) differentiation.
   Answer: B

32. The body cavity containing the urinary bladder and portions of the large intestine is the
   A) pelvic cavity.
   B) abdominal cavity.
   C) mediastinum.
   D) pleural cavity.
   E) dorsal cavity.
   Answer: A

33. The pericardial cavity contains the
   A) lungs.
   B) thyroid glands.
   C) brain.
   D) heart.
   E) stomach.
   Answer: D

34. Which of the following does NOT belong to the nine-abdominopelvic regions?
   A) left lumbar region
   B) right upper quadrant
   C) right iliac region
   D) epigastric region
   E) hypogastric region
   Answer: B

35. The majority of the stomach is found in the
    A) right hypochondriac region.
    B) left hypochondriac region.
    C) epigastric region.
    D) hypogastric region.
    E) umbilical region.
    Answer: C

36. Failure of the body to maintain homeostasis will
    A) have no effect on health.
    B) cause illness.
    C) always causes death.
    D) initiate positive feedback.
    E) increase body temperature.
    Answer: B

37. All of the following are controlled by homeostatic mechanisms EXCEPT
    A) muscle movement.
    B) blood pressure.
    C) body temperature.
    D) blood sugar levels.
    E) breathing rate.
    Answer: A

38. Which of the following processes is controlled by positive feedback?
    A) blood sugar levels
    B) blood clotting
    C) blood pressure
    D) body temperature
    E) breathing rate
    Answer: B

39. Collectively the organs are called
    A) mediastinum.
    B) mammary glands.
    C) abdominal organs.
    D) viscera.
    E) pleural organs.
    Answer: D

40. The basic structural and functional units of an organism are
    A) atoms.
    B) molecules.
    C) cells.
    D) tissues.
    E) organs.
    Answer: C

41. Atoms combine to form
    A) cells.
    B) organs.
    C) organ systems.
    D) molecules.
    E) None of the above are correct.
    Answer: D

42. The organ system which contains the skin and its derivatives is the
    A) skeletal system.
    B) integumentary system.
    C) muscular system.
    D) endocrine system.
    E) cardiovascular system.
    Answer: B

43. The anatomical term used to describe the region of the neck is
    A) facial.
    B) brachial.
    C) cervical.
    D) thoracic.
    E) mental.
    Answer: C

44. The splitting of proteins into amino acids is an example of
    A) anabolism.
    B) catabolism.
    C) metabolism.
    D) differentiation.
    E) homeostasis.
    Answer: B

45. Petunia Pig is trying on her new two-piece lingerie (trying to impress Porky, I think). Speaking anatomically, we could say her bra was _______ to her "skivvies" (bottoms).
    A) anterior
    B) medial
    C) superior
    D) inferior
    E.) distal
    Answer: C

46. The chin is superior to the
    A) nose.
    B) navel.
    C) forehead.
    D) eyebrows.
    E) lips.
    Answer: B

True-False

Write T if the statement is true and F if the statement is false.

1. Anatomy refers to the structure and function of the body.
   Answer: False

2. The chemical level of structural organization includes all chemicals needed to maintain life.
   Answer: True

3. The integumentary system protects all body systems.
   Answer: True

4. Anabolism is a chemical process, which does not require energy.
Answer: False

5. Reproduction refers to the formation of new cells for growth only.
Answer: False

6. Stress has an adverse effect on homeostasis because it creates an imbalance in the internal environment.
Answer: True

7. The control center determines the level at which the controlled condition needs to be maintained.
Answer: True

8. To describe the relationship of different body structures to each other, anatomists use directional terms.
Answer: True

9. The sagittal plane divides the body into equal superior and inferior portions.
Answer: False

10. The spinal cord and the brain are located in the cranial cavity.
Answer: False

11. The heart is located in the mediastinum.
Answer: True

12. For practical purposes clinicians prefer to use the quadrant division of the abdominopelvic region.
Answer: True

13. Childbirth is a good example of a positive feedback mechanism.
Answer: True

14. The heart is superior to the cranium.
Answer: False

15. A good example of an anabolic reaction would be the body digesting starches into smaller sugars.
Answer: False

16. Temperature regulation is a good example of negative feedback.
Answer: True

17. The dorsal cavity consists of the cranial and spinal cavities.
Answer: True

18. The following are in order from smallest to largest: Atom, cell, molecule, organ, tissue, organism.
Answer: False

19. The basic living structural and functional units of an organism are the cells.
Answer: True

Short Answer

Write the word or phrase that best completes each statement or answers the question.

1. The study dealing with the structures of the human body is called _______________.
   Answer: anatomy

2. All the body systems combined make up a(n) _________________.
   Answer: organism

3. Molecules combine to form the ____________ level of organization.
   Answer: cellular

4. The highest level of organization is the _____________________.
   Answer: organismic level

5. The ability to detect and respond to changes in the environment is called ___________________.
   Answer: responsiveness

6. The maintenance of relatively stable conditions for the cells of the human body is called _____________________.
   Answer: homeostasis

7. The process by which unspecialized cells become specialized cells is called _______.
   Answer: differentiation

8. The basic component of a feedback system that produces a response is the _________.
   Answer: effector

9. The component of a feedback system that monitors changes in the controlled condition is the ________________.
   Answer: receptor

10. The plane that divides the body into anterior and posterior positions is the _________ plane.
    Answer: frontal

11. The stomach is __________ to the lungs.
    Answer: inferior

12. The molecules produced and released by the endocrine system are referred to as _________.
    Answer: hormones

13. The humerus (upper arm) is _________ to the radius in the forearm.
    Answer: proximal

14. The _______________ cavity contains the heart.
    Answer: pericardial

15. The thoracic and abdominopelvic cavities are subdivisions of the _______________ body cavity.
    Answer: ventral

16. The anatomical term for the head is ________________.
    Answer: cephalic

17. All the bones of the body, their associated cartilages, and joints belong to the __________________ system.
    Answer: skeletal

18. Hair and nails belong to the ________ system.
    Answer: integumentary

19. Blood clotting is an example of ______________ feedback.
    Answer: positive

20. Homeostasis is mainly controlled by the ________ and __________ systems.
    Answer: nervous and endocrine

21. The basic structural and functional units of an organism are __________.
    Answer: cells

Matching

Choose the item from Column 2 that best matches each item in Column 1.

Match each term in Column 2 with its definition in Column 1.

1. Column 1: The study of the structure of the human body
   Column 2: Anatomy

2. Column 1: The study of the function of the human body
   Column 2: Physiology

3. Column 1: The lowest level of structural organization
   Column 2: Chemical level

4. Column 1: The sum total of all chemical reactions in the human body.
   Column 2: Metabolism

5. Column 1: A chemical reaction that uses energy.
   Column 2: Anabolism

Choose the correct anatomical term from Column 2 for each of the common names in Column 1.

6. Column 1: eye
   Column 2: orbital

7. Column 1: hand
   Column 2: manual

8. Column 1: cheek
   Column 2: buccal

9. Column 1: buttock
   Column 2: gluteal

10. Column 1: spinal column
    Column 2: vertebral

Essay

Write your answer in the space provided or on a separate sheet of paper.

1. Name and briefly describe the levels of structural organization in the human body.

   Answer: The chemical level—includes atoms and molecules.
   The cellular level—includes all different cells made of combinations of molecules.
   The tissue level tissues—consist of groups of similar cells.
   The organ level—organs are formed when different types of tissues join together.
   The system level—consists of related organs that have a common function.
   The organismic level—the highest level of structural organization includes all organ systems making up the entire organism.

2. Name and briefly describe the planes that can be passed through the human body.

   Answer: A sagittal plane divides the body into right and left portions.
   A midsagittal plane divides the body into equal right and left portions.
   A parasagittal plane divides the body into unequal right and left portions.
   A frontal (coronal) plane divides the body into anterior and posterior portions.
   A transverse plane divides the body into superior and inferior portions.
   An oblique plane passes through the body in an angle between the planes mentioned above.

3. Define homeostasis.
   Answer: Homeostasis is a condition in which the internal environment of the body is maintained within certain physiological limits.

CHAPTER 2 Introductory Chemistry

Multiple-Choice

Choose the one alternative that best completes the statement or answers the question.

1. Negatively charged particles in an atom are called
   A) neutrons.
   B) electrons.
   C) protons.
   D) elements.
   E) isotopes.
   Answer: B

2. When two or more atoms combine, the resulting combination is called
   A) atomic mass.
   B) atomic number.
   C) inert element.
   D) molecule.
   E) salt.
   Answer: D

3. The chemical symbol for sodium is
   A) K.
   B) O.
   C) Mg.
   D) Na.
   E) Sn.
   Answer: D

4. Substances that cannot be broken down into simpler substances by ordinary chemical reactions are called
   A) matter.
   B) compounds.
   C) chemical elements.
   D) inorganic molecules.
   E) organic molecules.
   Answer: C

5. The number of protons in an atom corresponds with the
   A) atomic mass.
   B) atomic number.
   C) number of neutrons and electrons.
   D) atomic weight.
   E) number of valences.
   Answer: B

6. An abundant element found in water and most organic molecules is
   A) nitrogen.
   B) hydrogen.
   C) potassium.
   D) carbon.
   E) sodium.
   Answer: B

7. The backbone of organic molecules is formed by
   A) carbon.
   B) oxygen.
   C) magnesium.
   D) phosphorus.
   E) nitrogen.
   Answer: A

8. A particle with a negative or positive charge is referred to as
   A) electron.
   B) neutron.
   C) ion.
   D) proton.
   E) isotope.
   Answer: C

9. A chemical bond where electrons are taken from one atom and given to another is a(n)
   A) hydrogen bond.
   B) ionic bond.
   C) single covalent bond.
   D) double covalent bond.
   E) polar covalent bond.
   Answer: B

10. An atom that gives up electrons is considered to be a(n)
    A) electron acceptor or anion.
    B) electron donor or cation.
    C) onion.
    D) electron.
    E) molecule.
    Answer: B

11. The bond which is found between water molecules such as in ice crystals is a(n)
    A) hydrogen bond.
    B) ionic bond.
    C) single covalent bond.
    D) double covalent bond.
    E) None of the above are correct.
    Answer: A

12. A chemical bond in which one pair of electrons is shared between atoms is a(n)
    A) hydrogen bond.
    B) ionic bond.
    C) single covalent bond.
    D) double covalent bond.
    E) triple covalent bond.
    Answer: C

13. The most common chemical bonds in the human body are
    A) covalent bonds.
    B) ionic bonds.
    C) hydrogen bonds.
    D) double bonds.
    E) None of the above.
    Answer: A

14. Which of the following best describes a synthesis reaction?
    A) Large molecules are broken down to form smaller ones.
    B) Molecules combine to form large molecules.
    C) A reaction that always requires chemical energy.
    D) A and C are correct.
    E) B and C are correct.
    Answer: E

15. All of the following are organic compounds EXCEPT
    A) nucleic acids.
    B) water.
    C) proteins.
    D) lipids.
    E) carbohydrates.
    Answer: B

16. Energy needed for chemical reactions in the body is provided by the breakdown of
    A) ribonucleic acid (RNA).
    B) deoxyribonucleic acid (DNA).
    C) adenosine diphosphate (ADP).
    D) adenosine triphosphate (ATP).
    E) adenosine monophosphate (AMP).
    Answer: D

17. In an average healthy adult, 55% to 60% of the body weight is composed of
    A) amino acids.
    B) salts.
    C) water.
    D) fat.
    E) proteins.
    Answer: C

18. All of the following are properties of water EXCEPT
    A) it is an excellent solvent.
    B) it absorbs heat very quickly.
    C) it can participate in chemical reactions.
    D) it serves as a lubricant.
    E) it releases heat very slowly.
    Answer: B

19. A molecule that dissociates in water and gives off hydrogen ions is a(n)
    A) salt.
    B) acid.
    C) base.
    D) buffer.
    E) solvent.
    Answer: B

20. The higher the number on the pH scale,
    A) the higher the hydrogen ion concentration.
    B) the higher the hydroxide ion concentration.
    C) the more acidic a solution.
    D) the more neutral a solution.
    E) the lower the hydroxide ion concentration.
    Answer: B

21. Which of the following describes the most acidic solution?
A) pH 4
B) pH 5
C) pH 7
D) pH 9
E) pH 14
Answer: A

22. The pH of blood is
A) 4-5.
B) 6.76-7.00.
C) 7.20-7.60.
D) 7.35-7.45.
E) 7.65-8.00.
Answer: D

23. To prevent drastic changes in the pH and to maintain homeostasis, the body
A) uses digestive enzymes.
B) increases the hydrogen ion concentration in the blood.
C) decreases the hydrogen ion concentration in the blood.
D) uses buffer systems.
E) changes the body temperature.
Answer: D

24. The storage form of glucose in the liver is
A) fructose.
B) glycogen.
C) fat.
D) starch.
E) glycerol.
Answer: B

25. The building blocks of carbohydrates are
A) polysaccharides.
B) disaccharide.
C) monosaccharides.
D) glycogen.
E) starches.
Answer: C

26. In humans, glycogen is stored in the cells of the
A) brain.
B) liver.
C) muscles.
D) Both A and B are correct.
E) Both B and C are correct.
Answer: E

27. Polysaccharides can be broken down into simpler sugars by the process of
A) dehydration synthesis.
B) simple synthesis.
C) hydrolysis.
D) anabolism.
E) None of the above are correct.
Answer: C

28. The most highly concentrated source of energy in the body is
   A) proteins.
   B) amino acids.
   C) glycogen.
   D) triglycerides.
   E) glucose.
   Answer: D

29. The building blocks of triglycerides are
   A) glycerol and fatty acids.
   B) chains of fatty acids.
   C) monosaccharides.
   D) chains of amino acids.
   E) nucleic acids.
   Answer: A

30. Chemically, certain sex hormones such as estrogens and testosterone are classified as
   A) proteins.
   B) carbohydrates.
   C) nucleic acids.
   D) lipids.
   E) starches.
   Answer: D

31. All of the following belong to the group of unsaturated fats EXCEPT
   A) palm oil.
   B) olive oil.
   C) sunflower oil.
   D) canola oil.
   E) corn oil.
   Answer: A

32. Chemically estradiol is a(n)
   A) amino acid.
   B) steroid.
   C) protein.
   D) enzyme.
   E) nucleic acid.
   Answer: B

33. The building blocks of proteins are
   A) carbons.
   B) nucleic acids.
   C) amino acids.
   D) glycerol.
   E) fatty acids.
   Answer: C

34. When three amino acids combine, the result is a(n)
   A) dipeptide.
   B) tripeptide.
   C) octapeptide.
   D) nonapeptide.
   E) polypeptide.
   Answer: B

35. The bonds formed between amino acids are
    A) peptide bonds.
    B) ionic bonds.
    C) hydrogen bonds.
    D) nitrogen bonds.
    E) None of the above occur between amino acids.
    Answer: A

36. All of the following statements about enzymes are true EXCEPT
    A) they speed up chemical reactions without changing themselves.
    B) enzymes are proteins.
    C) they are highly specific.
    D) they are biological catalysts.
    E) they are used up in a chemical reaction.
    Answer: E

37. All of the following are enzymes EXCEPT
    A) lactose.
    B) peptidase.
    C) oxidase.
    D) aminopeptidase.
    E) amylase.
    Answer: A

38. The building blocks of nucleic acids are
    A) amino acids.
    B) fatty acids.
    C) ribonucleic acids.
    D) nucleotides.
    E) peptides.
    Answer: D

39. Which of the following molecules contain the genetic information in a cell?
    A) ATP
    B) RNA
    C) DNA
    D) ADP
    E) mRNA
    Answer: C

40. Which of the following molecules contains the sugar ribose?
    A) RNA
    B) DNA
    C) ATP
    D) ADP
    E) AMP
    Answer: A

41. Which of the following molecules is considered to be a double helix?
    A) mRNA
    B) tRNA
    C) RNA
    D) DNA
    E) ATP
    Answer: D

42. Energy required to make ATP is supplied by a process called cellular
A) transcription.
B) dehydration.
C) translation.
D) respiration.
E) reproduction.
Answer: D

43. Which of the following is NOT a base found in DNA?
A) Guanine
B) Thymine
C) Uracil
D) Cytosine
E) Adenine
Answer: C

44. The double helix of DNA is stabilized by _______ bonds.
A) peptide bonds
B) hydrogen bonds
C) ionic bonds
D) polar covalent bonds
E) none of the above
Answer: B

45. The atomic weight of an atom is determined by
A) adding the total number of protons, neutrons and electrons.
B) determining the number of protons in the outer shell
C) adding the total number of neutrons, protons and proteins
D) adding the protons only
E) adding the total number of protons and neutrons.
Answer: E

46. Once a protein loses its normal configuration and is no longer able to perform its normal functions because of a disruption due to heat, chemicals, pH or radiation), this non working protein is said to be
A) deaminized.
B) diseased.
C) dysfunctional.
D) denatured.
E) dyslexic
Answer: D

47. DNA and RNA are composed of:
A) huge polymers of nucleotides.
B) small polypeptides of amino acids.
C) large polysaccharide chains.
D) unsaturated fatty acids.
E) phospholipids bilayers.
Answer: A

48. What is the smallest complete unit of an element?
A) atom
B) nucleus
C) electron shell
D) periodic table
E) matter
Answer: A

49. What is found in the atomic nucleus?
A) protons only
B) protons and neutrons
C) protons, neutrons, and electrons
D) DNA & RNA
E) none of the above
Answer: B

50. Which of the following is a synthesis reaction?
A) AB + CD → AC + BD
B) A + B → AB
C) AB → A + B
D) Acid + Base → salt + water
E) None of the above exemplify a synthesis reaction
Answer: B

True-False

Write T if the statement is true and F if the statement is false.

1. Chemical elements present in high concentrations in the human body are trace elements.
Answer: False

2. An element is composed of the same type of atoms.
Answer: True

3. Each electron shell of an atom can hold six electrons.
Answer: False

4. A compound is a chemical composed of two or more different elements.
Answer: True

5. Positively charged ions are called anions.
Answer: False

6. A covalent bond is more stable and harder to break than a hydrogen bond.
Answer: True

7. Decomposition reactions break down chemical bonds.
Answer: True

8. Inorganic compounds always contain carbon.
Answer: False

9. The combination of a solvent and a solution is called a solute.
Answer: False

10. Water requires a large amount of heat to change from a liquid to a gas.
Answer: True

11. A solution with a pH of 7 is slightly basic.
Answer: False

12. Carbohydrates are composed of carbon, hydrogen, and oxygen.
Answer: True

13. Glycogen is a monosaccharide.
    Answer: False

14. Cholesterol is used to produce some sex hormones.
    Answer: True

15. A change in the shape of a protein is called denaturation.
    Answer: True

16. Electrons are located in the atoms last shell only.
    Answer: False

17. Hydrogen bonding is one of the special properties that water and other molecules have that makes life possible on Earth.
    Answer: True

18. The pH scale is a logarithmic scale that represents hydrogen ion concentrations in a solution.
    Answer: True

19. Organic compounds always contain carbon and hydrogen.
    Answer: True

20. An inorganic compound can release electrolytes in water.
    Answer: True

21. Bases release OH- ions in solution.
    Answer: True

22. Organic compounds may contain hydrogen and carbon atoms but not both at the same time.
    Answer: False

Short Answer

Write the word or phrase that best completes each statement or answers the question.

1. Anything that occupies space and has mass is considered to be ____.
   Answer: matter

2. The portion of an atom which contains protons and neutrons is the ____.
   Answer: nucleus

3. The total number protons and neutrons in an atom is its ____.
   Answer: mass number

4. Atoms, which have a completely filled outer electron shell, are called _____.
   Answer: inert elements

5. Positively charged atoms are called ____.
   Answer: cations

6. An electrically charged ion or molecule that has an unpaired electron in its outermost shell is a(n) ____________.
   Answer: free radical

7. Substances in the cell that can combine with free radicals are called ___________.
   Answer: antioxidants

8. Atoms that pick up electrons from other atoms are called _____.
   Answer: electron acceptors

9. In covalent bonds, the electrons of two atoms are ____.
   Answer: shared

10. A chemical reaction in which a molecule is split apart is referred to as a(n) ____________ reaction.
    Answer: decomposition

11. Synthesis reactions are chemical reactions that ____ energy.
    Answer: require

12. The capacity to do work is called ____________.
    Answer: energy

13. A substance that can be dissolved in a solvent is called a ____.
    Answer: solute

14. Another term for "water loving" is ____________.
    Answer: hydrophilic

15. When salts dissolve in water they undergo _____.
    Answer: ionization (dissociation)

16. To describe the acidity or alkalinity of a solution, the _____scale can be used.
    Answer: pH

17. The higher the hydrogen ion concentration, the more ____ the solution.
    Answer: acidic

18. Monosaccharides and disaccharides are referred to as ____ sugars.
    Answer: simple

19. When many monosaccharides are joined together through dehydration synthesis a _____ is formed.
    Answer: polysaccharide

20. A fat in which all the carbon atoms are bonded to hydrogen atoms is called ____ fat.
    Answer: saturated

21. When two amino acids join together they form a ____.
    Answer: dipeptide

22. The portion of an enzyme where the substrate binds is called the ____.
    Answer: active site

23. Substances that speed up chemical reactions but do not change themselves are called _______.
    Answer: catalysts (enzymes)

24. The nitrogen base in DNA that always pairs with adenine is ____.
    Answer: thymine

25. When a phosphate group is removed from ATP, the molecule ____ is formed.
    Answer: ADP

Matching

Choose the item from Column 2 that best matches each item in Column 1.

Match each term in Column 2 with its definition in Column 1.

1. Column 1: Negatively charged particles of an atom.
   Column 2: Electron

2. Column 1: The smallest unit of matter.
   Column 2: Element

3. Column 1: A negatively charged ion.
   Column 2: Anion

4. Column 1: The number of deficient electrons in the outermost shell.
   Column 2: Valence

Match the chemicals in Column 1 with the group of organic compounds in Column 2.

5. Column 1: glycogen
   Column 2: carbohydrate

6. Column 1: steroids
   Column 2: lipid

7. Column 1: enzyme
   Column 2: protein

8. Column 1: RNA
   Column 2: nucleic acid

9. Column 1: prostaglandin
   Column 2: lipid

10. Column 1: fructose
    Column 2: carbohydrate

Essay

Write your answer in the space provided or on a separate sheet of paper.

1. Name the different chemical bonds and briefly describe how they are formed.
   Answer: Covalent bonds are formed by the sharing of one, two, or three pairs of electrons.
   Ionic bonds are formed when an actual transfer of electrons occurs between atoms.
   Hydrogen bonds are the weakest of the chemical bonds and are formed due to an attraction of a hydrogen atom of one molecule and an oxygen or nitrogen atom of another molecule.

2. Briefly explain the importance of water in the human body.
   Answer: Water makes up about 55 to 60 percent of body weight and therefore is the most abundant substance in the body. Water is an excellent solvent and suspending medium, it absorbs and releases heat very slowly, helping to maintain the homeostasis of body temperature.

3. Explain the difference between acids, bases, and salts.
   Answer: Acids are compounds, which dissociate in water and give off hydrogen ions. Bases give off hydroxide ions when they dissociate in water, and salts ionize in water into cations and anions, neither of which are hydrogen or hydroxide ions.

CHAPTER 3 Cells

Multiple-Choice

Choose the one alternative that best completes the statement or answers the question.

1. The plasma membrane consists of
   A) cellulose and carbohydrates.
   B) proteins mostly.
   C) entirely of phospholipids.
   D) phospholipids, proteins, and carbohydrates.
   E) carbohydrates and lipids.
   Answer: D

2. The cytoplasm is the term for
   A) all cell organelles combined.
   B) microtubules and microfilaments.
   C) the fluid portion of the cell.
   D) the cytosol plus cell organelles.
   E) the communication center of the cell.
   Answer: D

3. Which of the following statements are true for the plasma membrane?
   A) It is selectively permeable.
   B) It contains glycoproteins.
   C) It contains cholesterol.
   D) A and B are correct.
   E) A, B, and C are correct.
   Answer: E

4. The lipid bilayer is permeable to all of the following substances EXCEPT
   A) amino acids.
   B) fat-soluble vitamins.
   C) steroids.
   D) oxygen.
   E) water.
   Answer: A

5. Endocytosis is an example of
   A) excretion.
   B) passive transport.
   C) active transport.
   D) facilitated diffusion.
   E) simple diffusion.
   Answer: C

6. Which of the following is necessary for diffusion to take place?
   A) a concentration gradient
   B) a selectively permeable membrane
   C) a hypertonic solution
   D) cellular energy
   E) All of the above.
   Answer: A

7. A red blood cell placed in a hypotonic solution
   A) loses water.
   B) gains water.
   C) neither gains nor loses water.
   D) shrinks.
   E) will not change shape.
   Answer: B

8. An isotonic solution for human red blood cells is a
   A) 10% NaCl solution.
   B) 2% saline solution.
   C) 0.9% saline solution.
   D) water.
   E) 0.9% glucose solution.
   Answer: C

9. Pinocytosis and phagocytosis are a function of the
   A) cytoplasm.
   B) plasma membrane.
   C) ribosomes.
   D) mitochondria.
   E) cell nucleus.
   Answer: B

10. The movement of molecules from area of low concentration to an area of high concentration requires
    A) cellular energy.
    B) facilitated diffusion.
    C) integral proteins.
    D) Both A and B.
    E) Both A and C.
    Answer: E

11. Certain white blood cells can destroy bacteria by the process of
    A) pinocytosis.
    B) phagocytosis.
    C) exocytosis.
    D) lysis.
    E) None of the above.
    Answer: B

12. Which of the following statements BEST describes active transport?
    A) a concentration gradient is needed
    B) requires a carrier
    C) requires cellular energy
    D) requires osmotic pressure
    E) both B and C
    Answer: E

13. Chromatin is found in the
    A) nucleus.
    B) nuclear pores.
    C) ribosomes.
    D) mitochondria.
    E) lysosomes.
    Answer: A

14. The packaging and sorting of proteins is the function of the
    A) endoplasmic reticulum.
    B) Golgi complex.
    C) mitochondria.
    D) lysosomes.
    E) nucleus.
    Answer: B

15. Steroid synthesis is the function of the
    A) Golgi complex.
    B) ribosomes.
    C) rough endoplasmic reticulum.
    D) smooth endoplasmic reticulum.
    E) mitochondria.
    Answer: D

16. Digestive enzymes are found in
    A) Golgi complexes.
    B) rough endoplasmic reticulum.
    C) smooth endoplasmic reticulum.
    D) lysosomes.
    E) mitochondria.
    Answer: D

17. Protein synthesis occurs in structures attached to the
    A) smooth endoplasmic reticulum.
    B) rough endoplasmic reticulum.
    C) mitochondria.
    D) Golgi complexes.
    E) lysosomes.
    Answer: B

18. Which of the following are considered the "powerhouses" of the cell?
    A) lysosomes
    B) ribosomes
    C) nucleoli
    D) mitochondria
    E) peroxisomes
    Answer: D

19. Which of the following belongs to the cytoskeleton?
    A) cytosol
    B) mitochondria
    C) microtubules
    D) centromere
    E) flagella
    Answer: C

20. The structure located near the nucleus, made of two cylindrical structures composed of clusters of microtubules is the
    A) nucleolus.
    B) centrosome.
    C) flagellum.
    D) microtubule.
    E) cilium.
    Answer: B

21. Which of the following cell organelles help to detoxify the blood in the liver?
   A) nucleus
   B) lysosomes
   C) rough ER
   D) smooth ER
   E) vacuoles
   Answer: D

22. The process by which worn-out cell organelles are digested is called
   A) autolysis.
   B) autoregulation.
   C) autophagy.
   D) lysis.
   E) phagocytosis.
   Answer: C

23. The human somatic cells contain ______ chromosomes.
   A) 46
   B) 23
   C) 43
   D) 24
   E) 54
   Answer: A

24. Which of the following is a component of RNA only?
   A) andenine
   B) cytosine
   C) guanine
   D) thymine
   E) uracil
   Answer: E

25. The anticodon is located on the
   A) mRNA.
   B) rRNA.
   C) tRNA.
   D) DNA.
   E) ribosome.
   Answer: C

26. The process of translation during protein synthesis takes place
   A) in the nucleus.
   B) in the nucleolus.
   C) in the cytoplasm.
   D) on the ribosomes.
   E) on the mitochondria.
   Answer: D

27. Amino acids that participate in protein synthesis are picked up in the cytosol by
   A) mRNA.
   B) rRNA.
   C) tRNA.
   D) ribosomes.
   E) ATP.
   Answer: C

28. A CGT base triplet on DNA is copied into mRNA as
A) CAT.
B) GCA.
C) GCU.
D) ACU.
E) TGC.
Answer: B

29. When particular protein is complete, synthesis is stopped by a special
A) anticodon.
B) amino acid.
C) stop codon.
D) start codon.
E) carrier protein.
Answer: C

30. All of the following are nucleotide bases in DNA molecules EXCEPT
A) adenine.
B) cytosine.
C) guanine.
D) thymine.
E) uracil.
Answer: E

31. A protein is defined by the sequence of its
A) fatty acids.
B) amino acids.
C) molecules.
D) atoms.
E) enzymes.
Answer: B

32. The cell division which produces two identical cells is called
A) somatic cell division.
B) meiosis I.
C) meiosis II.
D) cytokinesis.
E) reproductive cell division.
Answer: A

33. The cytoplasmic division is referred to as
A) meiosis.
B) somatic cell division.
C) reproductive cell division.
D) mitosis.
E) cytokinesis.
Answer: E

34. The replication of DNA takes place during
A) mitosis.
B) meiosis I.
C) meiosis II.
D) interphase.
E) cytokinesis.
Answer: D

35. The assembly of microtubules that is responsible for the movement of chromosomes is the
A) centromere.
B) chromosome.
C) chromatin.
D) basal body.
E) mitotic spindle.
Answer: E

36. The splitting and separation of centromeres occurs during
A) prophase.
B) anaphase.
C) metaphase.
D) telophase.
E) cytokinesis.
Answer: B

37. The final stage of mitosis is
A) prophase.
B) interphase.
C) anaphase.
D) telophase.
E) metaphase.
Answer: D

38. The branch of medicine dealing with medical issues of aging is called:
A) elderly physiology.
B) gerontology.
C) oncology.
D) psychology.
E) geriatrics.
Answer: E

39. The study of tumors is called
A) pathology.
B) oncology.
C) epidemiology.
D) pharmacology.
E) histology.
Answer: B

40. The spread of cancerous cells to parts of the body is referred to as
A) hyperplasia.
B) metastasis.
C) malignancy.
D) mutation.
E) hypertrophy.
Answer: B

41. Any cancer causing agent is a(n)
A) oncogen.
B) mutant.
C) carcinogen.
D) neoplasm.
E) chemical.
Answer: C

42. All of the following belong to abnormal cell division EXCEPT
A) tumor.
B) hyperplasia.
C) neoplasm.
D) cleavage furrow.
E) benign tumor.
Answer: D

43. Meiosis is the nuclear division used during the formation of
A) sperm cells.
B) skin cells.
C) red blood cells.
D) cancer cells.
E) All of the above are correct.
Answer: A

44. Which of the following is only true of active transport and never true of passive transport?
A) can move substances from low to high concentrations
B) requires energy in the form of ATP
C) requires a carrier protein
D) only A & B are true of active transport
E) All of the above are true of active transport and not passive.
Answer: E

45. The major function of the ribosomes is to:
A) coordinate lipid synthesis.
B) coordinate protein synthesis.
C) detoxify drugs and alchohol.
D) act as the cell's garbage disposer or digestive system.
E) form vesicles.
Answer: B

46. A cell in a hypotonic environment will cause the following condition:
A) water molecules will move into a cell along a concentration gradient.
B) water molecules will move out of a cell along a concentration gradient.
C) water molecules wiill move into a cell against a concentration gradient.
D) water molecules will move out of a cell against a concentration gradient.
E) both A & D could happen.
Answer: A

47. All of the following are true about the cytoplasm of the cell except
A) site of chemical reactions in the cell.
B) responsible for nuclear division during mitosis.
C) has liquid portion known as the cytosol.
D) is composed of a lattice of protein filaments called the cytoskeleton.
E) contains the cell organelles.
Answer: B

48. Active transport requires proteins (often called pumps) because they
A) move molecules along a concentration gradient.
B) are necessary for cell metabolism.
C) are required for osmosis.
D) require no energy from the cell.
E) move molecules against the concentration gradient.
Answer: E

49. The cell organelle responsible for transferring food energy into ATP energy is the
   A) ribosome.
   B) mitochondira.
   C) lysosome.
   D) Golgi Apparatus.
   E) nuclear envelope.
   Answer: B

50. N. Ticing Scent loves perfume. She visits her local discount store regularly. Today she came to class wearing her latest purchase and soon everyone was familiar with the scent. What is being demonstrated here?
   A) osmosis
   B) diffusion
   C) phagocytosis
   D ctive transport
   E) wise shopping
   Answer: B

True-False

Write T if the statement is true and F if the statement is false.

1. The term cytoplasm refers to all cellular contents between the plasma membrane and the nucleus.
   Answer: True

2. Peripheral proteins penetrate all the way through the phospholipid bilayer of the plasma membrane.
   Answer: False

3. Most proteins can easily pass through the plasma membrane.
   Answer: False

4. Fluid outside the cells of the body is called intracellular fluid.
   Answer: False

5. Kinetic energy is the energy of motion of molecules.
   Answer: True

6. Facilitated diffusion requires a membrane carrier and cellular energy.
   Answer: False

7. Osmotic pressure is the pressure needed to stop water movement across membranes.
   Answer: True

8. An eight percent NaCl solution is isotonic to most cells.
   Answer: False

9. The nucleus can communicate with the cytosol via nuclear pores.
   Answer: True

10. Ribosomes consist of three subunits of equal size.
   Answer: False

11. The Golgi complex produces lysosomes.
   Answer: True

12. Multiple projections on the surface of cells that move the mucus in the respiratory tract are called flagella.
Answer: False

13. Human skin cells contain 23 chromosomes.
Answer: False

14. The new DNA after replication contains one old and one new strand.
Answer: True

15. The first stage of mitosis is called prophase.
Answer: True

16. Cytosine is found in both DNA and RNA and it only bonds with the base guanine.
Answer: True

17. Protein synthesis begins with a DNA code in the nucleus. DNA is copied as an mRNA strand. The mRNA goes to the ribosome where the tRNA brings in the appropriate amino acids. These amino acids bond to form a protein. The process continues until the stop code is read.
Answer: True

18. In passive transport, the net result is molecules move from a lower to a higher concentration.
Answer: False

19. Passive transport requires a cell's energy but active transport does not.
Answer: False.

Short Answer

Write the word or phrase that best completes each statement or answers the question.

1. The branch of science dealing with the study of cellular structure is _____ .
Answer: cytology

2. The basic chemical framework of the plasma membrane is the __________ __________.
Answer: lipid bilayer

3. Proteins loosely attached to the interior surface of the cell membrane are called ___________ proteins.
Answer: peripheral

4. The fluid in the human body that is contained inside the cells is called _____________ fluid.
Answer: intracellular

5. When molecules move from an area of high concentration to an area of low concentration, they move with the ____.
Answer: concentration gradient

6. When a solution has the same concentration of water molecules and solutes as a red blood cell, it is considered a(n) ____ solution.
Answer: isotonic

7. The shrinkage that occurs when cells are placed into a hypertonic solution is called ____.
Answer: crenation

8. The movement of water through a selectively permeable membrane is __________.
Answer: osmosis

9. Molecules are transported from an area of lower to an area of higher concentration by ____.
Answer: active transport

10. The projections of the plasma membrane formed by phagocytes around large solid particles outside the cell are ____.
Answer: pseudopods

11. The thick, semifluid portion of the cytoplasm is the ____.
Answer: cytosol

12. The nucleus is enclosed by the ____.
Answer: nuclear envelope (membrane)

13. All proteins that are exported from the cell are released to the exterior of the cell by the process of ___.
Answer: exocytosis

14. Cristae are a series of folds found in ____.
Answer: mitochondria

15. Microfilaments, microtubules, and intermediate filaments are structures of the ____.
Answer: cytoskeleton

16. Centrioles are cylindrical structures, which reside within the ____.
Answer: centrosome

17. A group of nucleotides on a DNA molecule that codes for a particular protein is called a(n) ____.
Answer: gene

18. A base triplet of AUG on mRNA would match the anticodon ____.
Answer: UAC

19. Anticodons are located on ____.
Answer: tRNA

20. The two major steps in protein synthesis are ____ and ____.
Answer: transcription, translation

21. Meiosis is also called the ____ cell division.
Answer: reproductive

22. The division of the parent cell's cytoplasm is called ____.
Answer: cytokinesis

23. A noncancerous growth is called a ____ tumor.
Answer: benign

24. A permanent structural change in a gene is a ____.
Answer: mutation

25. Compounds that appear to prevent cellular damage associated with cancer, aging, and heart disease are called ________________.
Answer: phytochemicals

Matching

Choose the item from Column 2 that best matches each item in Column 1.

1. Column 1: A disease that affects the entire body
   Column 2: systemic disease

2. Column 1: The science that deals with why, when, and where diseases occur.
   Column 2: epidemiology

3. Column 1: The study that deals with the effects and uses of drugs.
   Column 2: pharmacology

4. Column 1: The removal of tissue from the living body for diagnosis
   Column 2: biopsy

5. Column 1: The spread of cancer to other body tissues
   Column 2: metastasis

6. Column 1: Increase in the size of cells without cell division
   Column 2: hypertrophy

7. Column 1: A decrease in the size of cells
   Column 2: atrophy

Essay

Write your answer in the space provided or on a separate sheet of paper.

1. Name five cell organelles and briefly describe the function of each.
   Answer: Golgi complex--sorts, modifies, and packages proteins and lipids: the Golgi complexes also form lysosomes.
   Ribosomes are the site of protein synthesis.
   Lysosomes contain digestive enzymes and digest cell substances and foreign microbes.
   Mitochondria are the site of ATP production.
   The nucleus is the control center of the cell.
   The rough endoplasmic reticulum is responsible for the production of proteins for export.
   The smooth endoplasmic reticulum is the site of lipid synthesis.

2. Draw and label a generalized animal cell. Include six structures.
   Answer: needs to be determined by the individual instructor

3. Define osmosis.
   Answer: Osmosis is the diffusion of water through a selectively permeable membrane.

4. Briefly describe the stages of mitosis
   Answer:
   1. Prophase: the chromatin condenses and shortens into visible chromosomes, the nucleoli disappear, and the nuclear membrane breaks down. The mitotic spindle is formed during this stage of mitosis.
   2. Metaphase: the chromosomes line up on the metaphase plate (equatorial plate).
   3. Anaphase: the centromeres split, the sister chromatids separate, and each daughter chromosome moves toward the opposite poles of the cell.

4. Telophase: chromosomal movement stops, microtubules disappear, a nuclear envelope reforms, and cytokinesis occurs.

CHAPTER 4 Tissues

Multiple-Choice

Choose the one alternative that best completes the statement or answers the question.

1. Which of the following tissues is avascular?
   A) connective tissue
   B) muscle tissue
   C) skeletal tissue
   D) epithelial tissue
   E) nervous tissue
   Answer: D

2. A scientist who examines tissue changes that might indicate disease is called
   A) histologist.
   B) radiologist.
   C) pathologist.
   D) cytologist.
   E) endocrinologist.
   Answer: C

3. The tissue lining body cavities is the
   A) epithelial tissue.
   B) connective tissue.
   C) skeletal tissue.
   D) muscle tissue.
   E) nervous tissue.
   Answer: A

4. The epithelial tissue which contains cells of different shapes and is capable of distention is
   A) simple columnar epithelium.
   B) pseudostratified epithelium.
   C) transitional epithelium.
   D) stratified cuboidal epithelium.
   E) squamous epithelium.
   Answer: C

5. Goblet cells are found in which of the following tissues?
   A) nervous tissue
   B) columnar epithelium
   C) cuboidal epithelium
   D) connective tissue
   E) none of the above
   Answer: B

6. The flat, single layered tissue which allows for diffusion to occur is
   A) pseudostratified epithelium.
   B) mesothelium.
   C) simple columnar epithelium.
   D) simple squamous epithelium.
   E) transitional tissue.
   Answer: D

7. Which of the following tissues is often ciliated?
   A) transitional epithelium
   B) connective tissue
   C) cartilage
   D) columnar epithelium
   E) squamous epithelium
   Answer: D

8. The different types of epithelia are named according to
   A) cell size.
   B) location.
   C) cell shape and arrangement of layers.
   D) cell shape and location in the body.
   E) location and function.
   Answer C

9. Tissue that functions in support and protection of body organs, stores energy, and provides immunity is
   A) epithelial tissue.
   B) connective tissue.
   C) nervous tissue.
   D) muscle tissue.
   E) osseous tissue.
   Answer: B

10. Which of the following epithelia function in absorption and secretion?
   A) squamous epithelium
   B) columnar epithelium
   C) cuboidal epithelium
   D) Both A and B
   E) Both B and C
   Answer: E

11. The main function of stratified squamous epithelium is
   A) secretion.
   B) absorption.
   C) diffusion.
   D) protection.
   E) distension.
   Answer: D

12. Epithelium that appears to have several layers of cells but does not, is classified as
   A) simple epithelium.
   B) stratified epithelium.
   C) transitional epithelium.
   D) pseudostratified epithelium.
   E) columnar epithelium.
   Answer: D

13. Which of the following tissues contains a large amount of matrix?
   A) connective tissue
   B) glial tissue
   C) epithelial tissue
   D) muscle tissue
   E) nervous tissue
   Answer: A

14. All of the following cells can be found in connective tissue EXCEPT
    A) fibroblasts.
    B) macrophages.
    C) plasma cell.
    D) mast cells.
    E) glial cells.
    Answer: E

15. Mucous secreting goblet cells are found in all of the following organ systems EXCEPT
    A) urinary system.
    B) endocrine system.
    C) digestive system.
    D) respiratory system.
    E) reproductive system.
    Answer: B

16 The fibers in the matrix of connective tissue are made of
    A) carbohydrates.
    B) proteins.
    C) nucleic acids.
    D) lipids.
    E) hyaluronic acid.
    Answer: B

17. Which of the following is responsible for waterproofing of the skin?
    A) hyaluronic acid
    B) chondroitin sulfate
    C) collagen
    D) keratin
    E) phospholipids
    Answer: D

18. A single layer of cuboidal cells lining a secretory duct would be classified as
    A) simple squamous epithelium.
    B) simple cuboidal epithelium.
    C) simple columnar epithelium.
    D) stratified cuboidal epithelium.
    E) stratified squamous epithelium.
    Answer: B

19. A basement membrane is always present
    A) between the epithelial cells.
    B) between the epithelium and the muscle tissue.
    C) between the epithelium and the connective tissue.
    D) on the apical surface of the epithelium.
    E) between the epithelium and the blood vessels.
    Answer: C

20. All of the following are secretory products of exocrine glands EXCEPT
    A) mucus.
    B) oil.
    C) digestive enzymes.
    D) hormones.
    E) saliva.
    Answer: D

21. The slippery, viscous substance that binds cells together and lubricates joints is
    A) hyaluronic acid.
    B) chondroitin sulfate.
    C) collagen.
    D) reticular connective tissue.
    E) elastin.
    Answer: A

22. Connective tissue fibers that can be stretched considerably without breaking are
    A) collagen fibers.
    B) reticular fibers.
    C) elastic fibers.
    D) glucoprotein fibers.
    E) chondroitin fibers.
    Answer: C

23. All of the following are classified as loose connective tissue EXCEPT
    A) areolar connective tissue.
    B) adipose tissue.
    C) reticular connective tissue.
    D) elastic connective tissue.
    E) Both C and D.
    Answer: E

24. To which of the following tissues does cartilage belong?
    A) epithelial tissue
    B) nervous tissue
    C) areolar tissue
    D) muscle tissue
    E) connective tissue
    Answer: E

25. The presence of chondrocytes and elastic fibers indicates
    A) dense connective tissue.
    B) hyaline cartilage.
    C) fibrocartilage.
    D) elastic cartilage.
    E) bone.
    Answer: D

26. What is the main function of fibrous connective tissue?
    A) absorption
    B) support
    C) contraction
    D) lining of body cavities
    E) secretion
    Answer: B

27. Blood belongs to which major tissue type?
    A) epithelial tissue
    B) connective tissue
    C) skeletal tissue
    D) muscle tissue
    E) nervous tissue
    Answer: B

28. Membranes that line body cavities that open directly to the exterior are
   A) serous membranes.
   B) synovial membranes.
   C) mucous membranes.
   D) parietal membranes.
   E) visceral membranes.
   Answer: C

29. Which of the following contain a large amount of elastic connective tissue?
   A) heart valves
   B) the periosteum
   C) dermis
   D) epidermis
   E) wall of arteries
   Answer: E

30 The portion of a membrane that is closest to the organ is the
   A) mucous portion.
   B) parietal portion.
   C) synovial portion.
   D) visceral portion.
   E) pleural portion.
   Answer: D

31. The cells of mature cartilage are
   A) osteoblasts.
   B) osteocytes.
   C) chondroblasts.
   D) chondrocytes.
   E) fibroblasts.
   Answer: D

32 The basic unit of compact bone is a(n)
   A) osteon.
   B) trabeculae.
   C) osteocyte.
   D) osteoblast.
   E) canaliculus.
   Answer: A

33. The matrix of blood is
   A) serum.
   B) plasma.
   C) water.
   D) intracellular fluid.
   E) lymph.
   Answer: B

34. Intercalated discs are structures of
   A) smooth muscle.
   B) epithelial tissue.
   C) cardiac muscle.
   D) skeletal muscle.
   E) connective tissue.
   Answer: C

35. All of the following are structures of nervous tissue EXCEPT
   A) neuroglia.
   B) axons.
   C) dendrite.
   D) cell body.
   E) periosteum.
   Answer: E

36. The membrane covering the lungs is the
   A) peritoneum.
   B) pleura.
   C) pericardium.
   D) mesothelium.
   E) endothelium.
   Answer: B

37. The membranes that line the cavities of some joints are
   A) synovial membranes.
   B) visceral membranes.
   C) mucous membranes.
   D) serous membranes.
   E) cutaneous membranes.
   Answer: A

38. The tissue that is highly specialized for contraction is
   A) nervous tissue.
   B) epithelial tissue.
   C) connective tissue.
   D) muscle tissue.
   E) osseous tissue.
   Answer: D

39. Neurolgia cells belong to which of the following tissues?
   A) epithelial tissue
   B) osseous tissue
   C) connective tissue
   D) muscle tissue
   E) nervous tissue
   Answer: E

40. Which of the following is an autoimmune disease?
   A) cystic fibrosis
   B) rheumatoid arthritis
   C) hypertension
   D) both A and B
   E) None of the above.
   Answer: A

41. What is the correct term for the serous membrane that acts as the inner lining of a cavity?
   A) parietal
   B) stratum basale
   C) visceral
   D) stratum corneum
   E) synovial
   Answer: A

42. Which has the best blood supply in the following list?
   A) epithelium
   B) loose or areolar connective tissue
   C) dense connective tissue
   D) ligaments
   E) hyaline cartilage
   Answer: B

43. What type of epithelial tissue forms the epidermal layer of the skin?
   A) simple squamous epithelium
   B) simple cuboidal epithelium
   C) stratified squamous epithelium
   D) stratified cuboidal epithelium
   E) simple columnar epithelium
   Answer: C

44. All of the following traits about connective and epithelial tissue are true <u>except</u>
   A) connective tissue cells are farther from each other than epithelial cells
   B) there is intercellular material or matrix between connective cells but usually not in epithelium.
   C) epithelial cells have a good blood supply that often diffuses into the connective tissue to nourish them as well.
   D) epithelial tissue cells can divide easily, often connective tissue repairs slowly or not at all.
   E) all of these statements about these tissues are true.
   Answer: C

45. Which type of connective tissue has the function of binding skin to organs and fills the spaces between muscles?
   A) loose connective tissue
   B) dense connective tissue
   C) hyaline cartilage
   D) fibrous cartilage
   E) elastic connective tissue.
   Answer: A

46 pithelial tissue has which of the following functions?
   A) secretion
   B) absorption
   C) protection
   D) lining spaces
   E) All of the above
   Answer: E

47. Which type of tissue is responsible for most secretions?
   A) nervous
   B) epithelium
   C) muscle
   D) connective
   E) None of the above.
   Answer: B

True-False

Write T if the statement is true and F if the statement is false.

1. A tissue is a group of similar cells that perform a specialized activity.
   Answer: True

2. The lining of body cavities is one of the functions of connective tissue.
Answer: False

3. Epithelial cells are arranged in continuous sheets.
Answer: True

4. The exchange of materials between epithelium and connective tissue is an active transport mechanism.
Answer: False

5. Squamous epithelium has cube-shaped cells.
Answer: False

6. Epithelial tissue has a high mitotic capacity.
Answer: True

7. Simple epithelium always consists of a layer of flat cells.
Answer: False

8. Pseudostratified epithelium is never ciliated.
Answer: False

9. Nonciliated columnar epithelium contains microvilli and goblet cells.
Answer: True

10. Stratified columnar epithelium is a common type of tissue in the human body.
Answer: False

11. Exocrine glands secrete their products into ducts.
Answer: True

12. Cartilage is highly vascular.
Answer: False

13. Connective tissue cells produce the ground substance of the connective tissue matrix.
Answer: True

14. Neuroglia protects and supports neurons.
Answer: True

15. Hyaline cartilage is found at the end of most long bones where joints form.
Answer: True

16. Epithelial tissue is specialized for protection, secretion, and absorption.
Answer: True

17. The difference between simple and stratified epithelium is in the shape of the cells.
Answer: False

18. The difference between dense and loose connective tissue is mainly in the arrangement of the fibers.
Answer: True

19. There are two basic types of nervous tissue cells, the neurons which support and the neuroglia which conduct impulses.
Answer: False

Short Answer

Write the word or phrase that best completes each statement or answers the question.

1. The science that deals with the study of tissues is called ____.
   Answer: histology

2. Epithelium that consists of two or more layers of cells is ____ epithelium.
   Answer: stratified

3. The epithelium lining the urinary bladder is ____ epithelium.
   Answer: transitional

4. The pituitary gland is an example of a(n) _____ gland.
   Answer: endocrine

5. Exocrine glands are made out of ________ tissue.
   Answer: epithelial

6. Blood vessels are lined by ___________.
   Answer: endothelium

7. A matrix and a cellular component are characteristics of ____ tissue.
   Answer: connective

8. The cells producing fibers in connective tissue are ____.
   Answer: fibroblasts

9. The most abundant connective tissue fibers are ____ fibers.
   Answer: collagen

10. Immature cells of cartilage are ____________.
    Answer: chondroblasts

11. Blood is classified as ____ tissue.
    Answer: connective

12. Cells that are specialized for fat storage are ____.
    Answer: adipocytes

13. A connective tissue where bundles of collagen fibers have an orderly parallel arrangement is classified as ____connective tissue.
    Answer: regular (dense regular)

14. The cells of mature cartilage are ____.
    Answer: chondrocytes

15. The type of cartilage that reduces friction at the joints is ____ cartilage.
    Answer: hyaline

16. Bone tissue is also called ____ tissue.
    Answer: osseous

17. The chemical that dilates small blood vessels and is released by mast cells is _________.
    Answer: histamine

18. The type of membrane in the body that does not contain epithelium is a(n) ____ membrane.
Answer: synovial

19. The part of a serous membrane attached to the wall of the body cavity is the ____ portion.
Answer: parietal

20. Synovial membranes are found in ____.
Answer: joints

21. Cells that are capable of generating and conducting nerve impulses are _____.
Answer: neurons (nerve cells)

22. The lining of the abdominal cavity is the _____.
Answer: peritoneum

23. The covering of the heart is the _____.
Answer: pericardium

24. The type of connective tissue that contains plasma is ____.
Answer: blood

25. The procedure developed to grow skin in the laboratory is called ______ ________.
Answer: tissue engineering

Matching

Choose the item from Column 2 that best matches each item in Column 1.

1. Column 1: A tissue that can store fat
Column 2: Connective tissue

2. Column 1: A tissue that is avascular
Column 2: Epithelium

3. Column 1: A group of specialized cells which secrete their products into ducts
Column 2: Exocrine glands

4. Column 1: Fibers that consist of collagen and glycoproteins
Column 2: Reticular fibers

5. Column 1: An embryonic connective tissue
Column 2: Mesenchyme

6. Column 1: A type of loose connective tissue
Column 2: Adipose tissue

7. Column 1: Lines body cavities that do not open to the exterior
Column 2: Serous membranes

Essay

Write your answer in the space provided or on a separate sheet of paper.

1. Name and briefly describe the function of the four major tissue types.
   Answer: 1. Epithelial tissue covers body surfaces and lines body cavities
   2. Connective tissue protects and supports the body and its organs, binds organs together, stores energy reserves, and provides immunity
   3. Muscular tissue is responsible for movement
   4. Nervous tissue initiates and transmits nerve impulses

2. Classify epithelia by shape and arrangement of cell layers.
   Answer: simple squamous epithelium
   simple cuboidal epithelium
   simple columnar epithelium
   stratified squamous epithelium
   stratified cuboidal epithelium
   stratified columnar epithelium
   transitional epithelium
   pseudostratified columnar epithelium

3. Distinguish between exocrine and endocrine glands
   Answer: Exocrine glands secrete their products into ducts, while endocrine glands secrete their products into the blood.

CHAPTER 5 The Integumentary System

Multiple-Choice

Choose the one alternative that best completes the statement or answers the question.

1. The skin belongs to the
   A) skeletal system.
   B) integumentary system.
   C) muscular system.
   D) sensory system.
   E) circulatory system.
   Answer: B

2. All of the following reflect homeostatic imbalances of the body manifested in the skin EXCEPT
   A) rashes.
   B) chicken pox.
   C) measles.
   D) cold sores.
   E) warts.
   Answer: E

3. The medical specialty that deals with the diagnosis and treatment of skin disorders is called
   A) dermatology.
   B) pathology.
   C) pharmacology.
   D) cardiology.
   E) oncology.
   Answer: A

4. The outermost portion of the skin is the
   A) hypodermis.
   B) subcutaneous layer.
   C) dermis.
   D) epidermis.
   E) basement membrane.
   Answer: D

5. All of the following are functions of the skin EXCEPT
   A) protection.
   B) vitamin B synthesis.
   C) excretion
   D) temperature regulation.
   E) sensation.
   Answer: B

6. The protein that helps protect and water proof the skin and underlying cells is
   A) melanin.
   B) melatonin.
   C) keratin.
   D) actin.
   E) carotene.
   Answer: C

7. The cells producing the pigment responsible for skin color are the
   A) keratinocytes.
   B) melanocytes.
   C) Langerhans cells.
   D) Merkel cells.
   E) adipocytes.
   Answer: B

8. Cells of the skin that are active in the immune process are
   A) keratinocytes.
   B) melanocytes.
   C) Merkel cells.
   D) adipocytes.
   E) Langerhans cells.
   Answer: E

9. The layer of the epidermis which contains cells capable of continued cell division is the
   A) stratum spinosum.
   B) stratum basale.
   C) stratum corneum.
   D) stratum granulosum.
   E) stratum lucidum.
   Answer: B

10. The layer of the epidermis primarily found in the thick skin of the palms and soles is the
    A) stratum basale.
    B) stratum spinosum.
    C) stratum granulosum.
    D) stratum lucidum.
    E) stratum corneum.
    Answer: D

11. The corpuscles of touch (Meissner's corpuscles) are found in the
    A) hypodermis.
    B) dermis.
    C) subcutaneous layer.
    D) epidermis.
    E) basement membrane.
    Answer: B

12. All of the following are pigments involved in skin color EXCEPT
    A) keratin.
    B) melanin.
    C) carotene.
    D) hemoglobin.
    E) All of the above.
    Answer: A

13. An inherited inability of an individual to produce melanin results in
    A) freckles.
    B) melanoma.
    C) erythema.
    D) albinism.
    E) carotenosis.
    Answer: D

14. The redness of the skin that can occur during allergic reactions is
    A) albinism.
    B) erythema.
    C) due to carotene.
    D) due to melanine.
    E) caused by hemoglobin.
    Answer: B

15. The primary function of hair is
    A) protection.
    B) sensory perception.
    C) to increase surface area.
    D) temperature regulation.
    E) movement
    Answer: A

16. The base of a hair follicle is enlarged into an onion-shaped structure called
    A) root.
    B) shaft.
    C) papilla.
    D) bulb.
    E) matrix.
    Answer: D

17. The bundle of smooth muscles associated with hair follicles is called
    A) orbicularis muscle.
    B) sebaceous bundle.
    C) arrector pili.
    D) goose bumps.
    E) none of the above.
    Answer: C

18. The glands usually associated with hair follicles are
    A) sudoriferous glands.
    B) sweat glands.
    C) ceruminous glands.
    D) eccrine glands.
    E) sebaceous glands.
    Answer: E

19. Glands that are present in the external auditory meatus are
    A) ceruminous glands.
    B) sudoriferous glands.
    C) sebaceous glands.
    D) eccrine glands.
    E) apocrine glands.
    Answer: A

20. Perspiration is the substance produced by
    A) ceruminous glands.
    B) sudoriferous glands.
    C) sebaceous glands.
    D) oil glands.
    E) holocrine glands.
    Answer: B

21. The portion of a nail that extends past the end of the finger is the
    A) nail body.
    B) nail root.
    C) lunula.
    D) free edge.
    E) nail matrix.
    Answer D

22. The cuticle of the nail consists of
    A) stratum corneum.
    B) stratum lucidum.
    C) nail matrix.
    D) nail root.
    E) basement membrane.
    Answer: A

23. All of the following are true for the aging skin EXCEPT
    A) collagen fibers decrease in number.
    B) elastic fibers loose some elasticity.
    C) macrophages increase in number.
    D) melanocytes increase in size.
    E) fibroblasts decrease in number.
    Answer: C

24. The layer of skin that contains dead cells filled with keratin is the
    A) stratum basale.
    B) stratum lucidum.
    C) stratum corneum.
    D) stratum spinosum.
    E) stratum granulosum.
    Answer: C

25. Receptors in the skin that are sensitive to touch are
    A) dermal papillae.
    B) chemoreceptors.
    C) Pacinian corpuscles.
    D) Meissner's corpuscles.
    E) C and D are correct
    Answer: D

26. A burn that involves the entire epidermis and some of the dermis is a
    A) first-degree burn.
    B) second-degree burn.
    C) third-degree burn.
    D) fourth-degree burn.
    E) blister.
    Answer: B

27. An inflammation of sebaceous glands is called
    A) dermatitis.
    B) decubitus.
    C) ulcer.
    D) acne.
    E) burn.
    Answer: D

28. The most common form of skin cancer is
    A) basal cell carcinomas.
    B) squamous cell carcinomas.
    C) granular cell carcinomas.
    D) malignant melanomas.
    E) epidermal melanomas.
    Answer: A

29. Which of the following skin infections is caused by a fungus?
    A) cold sores
    B) fever blisters
    C) hives
    D) Athlete's foot
    E) cysts
    Answer: D

30. If the arector pili muscles stay contracted for several minutes, we would assume the individual
    A) is having cramps.
    B) is having a tetanus infection
    C) is cold.
    D) is relaxed.
    E) is in love.
    Answer: C

31. Hair and skin color depend on:
    A) cerumen.
    B) keratin.
    C) melanin.
    D) sebum.
    E) Clairol #34.
    Answer: C

32. Where are the melanocytes found?
    A) parietal layer
    B) stratum basale
    C) visceral layer.
    D) stratum corneum
    E) stratum lucidum
    Answer: B

33. Which layer of the skin becomes keratinized.?
    A) parietal layer
    B) stratum basale
    C) stratum lucidum
    D) stratum corneum
    E) dermis.
    Answer: D

34. Which layer contains blood vessels?
    A) epidermis
    B) stratum basale
    C) stratum corneum
    D) dermis
    E) parietal layer
    Answer: D

35 If you were on a frantic search for adipose tissue, you would most likely find it in what skin layer?
A) dermis
B) epidermis
C) the keratinized layers
D) subcutaneous layer
E) anywhere on your instructor's body
Answer: D

36. Which layer of the skin is associated with protection from the ultraviolet light?
A) epidermis
B) dermis
C) subcutaneous layer
D) basement membrane
E) parietal layer
Answer: A

37. Which of the following is not correct concerning the skin?
A) The dermis is usually thicker than the epidermis.
B) The stratum basale forms the deepest layer of the epidermis.
C) The subcutaneous layer is between the dermis and epidermis.
D) The dermis contains smooth muscle and nerve tissue.
E) The epidermis is avascular.
Answer: C

True-False

Write T if the statement is true and F if the statement is false.

1. The hypodermis is the layer of skin below the epidermis.
Answer: False

2. Langerhans cells are important components in the immune system.
Answer: True

3. The skin produces Vitamin D after exposure to UV light.
Answer: True

4. The epidermis is the outermost layer of the skin.
Answer: True

5. The stratum spinosum of the skin has a high mitotic activity.
Answer: False

6. The dermis contains both elastic and collagen fibers.
Answer: True

7. The lower layer of the epidermis contains adipose tissue.
Answer: False

8. Carotene is a pigment of the skin.
Answer: True

9. Carotene tends to form patches called freckles.
Answer: False

10. Scalp hair grows continually throughout life.
    Answer: False

11. The matrix produces new hairs by cell division.
    Answer: True

12. Dark colored hair contains mostly true melanin.
    Answer: True

13. Sebum is secreted by sweat glands.
    Answer: False

14. The arector pili muscle attaches to the hairs' root.
    Answer: False

15. Wrinkles are due to decrease of elasticity in elastic fibers of the skin.
    Answer: False

16. The best place to find adipose tissue in the skin is in the dermis.
    Answer: False

17. Ceruminous glands are found in the external auditory canal and secrete wax.
    Answer: True

18. Freckles are an accumulation of melanin.
    Answer: True

Short Answer

Write the word or phrase that best completes each statement or answers the question.

1. The pigment found in specialized cells of the stratum basale is ____.
   Answer: melanin

2. Cold receptors are found in the ____.
   Answer: dermis

3. Cancer of ____ is considered to be one of the most serious skin cancers.
   Answer: melanocytes

4. The superficial portion of the hair is the ____.
   Answer: shaft

5. Plates of tightly packed, hard, keratinized cells of the epidermis are referred to as ____.
   Answer: nails

6. The portion of the nail that is visible is the ____.
   Answer: nail body

7. The whitish portion of the nail is the ____.
   Answer: lunula

8. A burn that involves the surface of the epidermis only is a ____ -degree burn.
   Answer: first

9. Bedsores are ____ sores.
Answer: pressure

10. Excessive exposure to the sun can result in ____.
Answer: skin cancer

11. Cold sores are caused by a ____.
Answer: virus

12. Medication that is applied to the skin surface is ____.
Answer: topical

13. The layer below the dermis is called ____.
Answer: hypodermis (subcutaneous layer)

14. Merkel cells are found in the ____.
Answer: epidermis

15. In the upper layer of the dermis, finger-like extensions called ____ are present to increase the surface area.
Answer: dermal papillae

16. Each hair follicle goes through a growth stage and a ____ stage.
Answer: resting

17. The color of blackheads is due to _____ and _____.
Answer: melanin, oxidized oil

18. Sweat glands associated with hair follicles are classified as ____ glands.
Answer: apocrine

Matching

Choose the item from Column 2 that best matches each item in Column 1.

1. Column 1: A portion of skin that has been scraped away
Column 2: Abrasion

2. Column 1: Caused by herpes simplex virus
Column 2: Cold sore

3. Column 1: Condition of the skin marked by redness and often-itchy patches
Column 2: Hives

4. Column 1: Removal of tattoos by a high-speed brush
Column 2: Dermabrasion

5. Column 1: A superficial skin infection, mostly in children, caused by staphylococci or streptococci
Column 2: Impetigo

6. Column 1: A patch called a birthmark
Column 2: Nevus

7. Column 1: Formation of a hardened growth of tissue
Column 2: Keratosis

8. Column 1: Mass produced by uncontrolled growth of epithelial cells due to a virus
   Column 2: Wart

9. Column 1: Wound or irregular tear of the skin
   Column 2: Laceration

10. Column 1: A superficial fungal infection of the skin of the foot
    Column 2: Athlete's feet

Essay

Write your answer in the space provided or on a separate sheet of paper.

1. Name the functions of the skin.
   Answer: 1. Regulation of body temperature
   2. Protection
   3. Sensation
   4. Excretion
   5. Immunity
   6. Vitamin D synthesis

2. Name the layers of the skin starting at the deepest to the most superficial layer.
   Answer: 1. Stratum basale
   2. Stratum spinosum
   3. Stratum granulosum
   4. Stratum lucidum
   5. Stratum corneum

3. Name the glands associated with the skin, and name the product each of these glands secrete.
   Answer: 1. Sebaceous glands (oil glands) secrete sebum, an oily substance
   2. Sudoriferous glands (sweat glands) secrete perspiration or sweat.
   3. Ceruminous glands secrete the combined product of sebaceous glands and ceruminous glands, which is called cerumen.

CHAPTER 6 The Skeletal System

Multiple-Choice

Choose the one alternative that best completes the statement or answers the question.

1. All of the following are functions of the skeletal system EXCEPT
   A) support.
   B) blood cell production.
   C) blood protein production.
   D) mineral storage.
   E) movement.
   Answer: C

2. Which of the following is NOT considered to be a long bone?
   A) bones of the ribcage
   B) bones of the thighs
   C) bones of the forearms
   D) bones of the fingers
   E) bones of the legs
   Answer: A

3. Which of the following are classified as flat bones?
   A) sesamoid bones
   B) vertebrae
   C) ankle bones
   D) cranial bones
   E) spongy bones
   Answer: D

4. Yellow bone marrow
   A) stores triglycerides.
   B) consists of mainly adipose cells.
   C) produces red blood cells.
   D) A and B are correct.
   E) A and C are correct.
   Answer: D

5. The shaft of a long bone is called the
   A) epiphysis.
   B) diaphysis.
   C) metaphysis.
   D) periosteum.
   E) endosteum.
   Answer: B

6. The site where bone growth occurs in a long bone is the
   A) epiphysial plate.
   B) epiphysis.
   C) diaphysis.
   D) articular cartilage.
   E) periosteum.
   Answer: A

7. The endosteum
   A) contains red bone marrow.
   B) is the unit of compact bone.
   C) lines the medullary cavity.
   D) surrounds the bone surface.
   E) is the end of long bones.
   Answer: C

8. Osteogenic cells
   A) are mature bone cells.
   B) develop into osteoblasts.
   C) break down bone.
   D) deposit calcium into the bone.
   E) surround themselves with matrix.
   Answer: B

9. Osteoclasts
   A) deposit calcium into bone.
   B) destroy bone matrix.
   C) are a type of white blood cell.
   D) B and C are correct.
   E) None of the above are correct.
   Answer: D

10. Nutrient arteries and nerves from the periosteum penetrate compact bone through the
    A) Haversian canal.
    B) central canal.
    C) Volkman's canal.
    D) concentric lamellae.
    E) canaliculi.
    Answer: C

11. Fat is stored in
    A) compact bone.
    B) spongy bone.
    C) red bone marrow.
    D) yellow bone marrow.
    E) articular cartilage.
    Answer: D

12. The secondary ossification center develops
    A) during intramembranous ossification.
    B) after development of the trabeculae.
    C) when blood vessels enter the epiphyses.
    D) at the appearance of the epiphyseal line.
    E) simultaneously with the primary ossification center.
    Answer: C

13. The cells responsible for the re-absorption of bone are
    A) osteoprogenitor cells.
    B) osteoclasts.
    C) osteocytes.
    D) osteoblasts.
    E) bone forming cells.
    Answer: B

14. The hormone(s) responsible for the homeostasis of blood calcium levels is (are)
    A) calcitonin.
    B) thyroid hormones.
    C) insulin and glucagon.
    D) parathyroid hormone and calcitonin.
    E) parathyroid hormone only.
    Answer: D

15. The rate of bone loss during aging can be slowed down by
    A) regular walking.
    B) rest and diet.
    C) diet only.
    D) vitamin injections.
    E) steroid treatment.
    Answer: A

16. All of the following hormones play a role in bone growth and maintenance EXCEPT
    A) human growth hormone.
    B) parathyroid hormone.
    C) insulin.
    D) thyroid hormones.
    E) epinephrine.
    Answer: E

17. Which of the following substances does NOT have a known effect on bone metabolism?
    A) vitamin D
    B) vitamin E
    C) vitamin A
    D) calcium
    E) phosphorus
    Answer: B

18. An opening in bone through which blood vessels and nerves pass is a
    A) depression.
    B) tuberosity.
    C) sinus.
    D) meatus.
    E) foramen.
    Answer: E

19. A depression in or on a bone is a
    A) fossa.
    B) tuberosity.
    C) foramen.
    D) condyle.
    E) meatus.
    Answer: A

20. The adult human skeleton consists of
    A) 230 bones.
    B) 200 bones.
    C) 206 bones.
    D) 198 bones.
    E) 302 bones.
    Answer: C

21. All of the following are facial bones EXCEPT
   A) nasal bone.
   B) maxilla.
   C) mandible.
   D) ethmoid bone.
   E) vomer.
   Answer: D

22. Which of the following is a cranial bone?
   A) vomer bone
   B) lacrimal bone
   C) zygomatic bone
   D) sphenoid bone
   E) palatine bone
   Answer: D

23. The suture between the parietal bones and the temporal bones is the
   A) sagittal suture.
   B) parasaggittal suture.
   C) lamboid suture.
   D) coronal suture.
   E) squamous suture.
   Answer: E

24. The temporal and zygomatic bones join to form the
   A) zygomatic arch.
   B) mandibular fossa.
   C) foramen magnum.
   D) mastoid process.
   E) styloid process.
   Answer: A

25. The medulla oblongata passes through the
   A) mental foramen.
   B) foramen magnum.
   C) occipital condyles.
   D) foramen ovale.
   E) carotid foramen.
   Answer: B

26. Which of the following forms the roof of the nasal cavity?
   A) superior nasal concha
   B) inferior nasal concha
   C) cribriform plate
   D) perpendicular plate
   E) nasal septum
   Answer: C

27. The upper jawbone is formed by the
   A) alveoli.
   B) maxillae.
   C) mandible.
   D) zygomatic bones.
   E) lacrimal bones.
   Answer: B

28. Which of the following is the only movable bone of the skull?
    A) maxilla
    B) zygomatic bone
    C) lacrimal bone
    D) mandible
    E) hyoid bone
    Answer: D

29. Technically, the cheekbones are called
    A) zygomatic bones.
    B) lacrimal bones.
    C) maxillae.
    D) vomer bones.
    E) frontal bones.
    Answer: A

30. The bone that does not articulate with any other bone is the
    A) vomer bone.
    B) zygomatic bone.
    C) lacrimal bone.
    D) ethmoid bone.
    E) hyoid bone.
    Answer: E

31. The adult vertebral column contains
    A) 32 vertebra.
    B) 30 vertebra.
    C) 26 vertebra.
    D) 24 vertebra.
    E) 20 vertebra.
    Answer: C

32. A spinal nerve leaves the vertebral column at the
    A) interventricular foramen.
    B) intervertebral foramen.
    C) vertebral arch.
    D) vertebral foramen.
    E) foramen magnum.
    Answer: B

33. Which of the following terms do NOT belong to a vertebrum?
    A) transverse process
    B) lateral process
    C) spinous process
    D) superior articular process
    E) pedicles
    Answer: B

34. The largest and strongest of the vertebrae in the spinal column are the
    A) cervical vertebrae.
    B) thoracic vertebrae.
    C) lumbar vertebrae.
    D) sacral vertebrae.
    E) coccygeal vertebrae.
    Answer: C

35. Which of the following vertebra articulate with the ribs?
A) sacral vertebra
B) lumbar vertebra
C) cervical vertebra
D) thoracic vertebra
E) coccygeal vertebra
Answer: D

36. All of the following are bones of the thorax EXCEPT
A) clavicle.
B) sternum.
C) true ribs.
D) thoracic vertebrae.
E) false ribs.
Answer: A

37. The pectoral girdle consists of the
A) humerus and radius.
B) ulna and radius.
C) hipbones.
D) sternum and ribs.
E) clavicle and scapula.
Answer: E

38. The bone of the upper arm is the
A) radius.
B) ulna.
C) humerus.
D) femur.
E) tibia.
Answer: C

39. All of the following are carpals EXCEPT
A) cuboid.
B) lunate.
C) hamate.
D) capitate.
E) scaphoid.
Answer: A

40. The anatomical term for shinbone is
A) fibula.
B) tibia.
C) humerus.
D) femur.
E) ulna.
Answer: B

41. The heaviest bone in the body is the
A) humerus.
B) fibula.
C) tibia.
D) hipbone.
E) femur.
Answer: E

42. Which of the following is NOT a bone of the foot?
    A) navicular
    B) calcaneus
    C) cuboid
    D) talus
    E) pisiform
    Answer: E

43. Which of the following has the function to build bone tissue?
    A) Osteoblast
    B) Osteoclast
    C) Osteocyte
    D) Osteomyte
    E) Chrondrocyte
    Answer: A

44. Long bones develop by
    A) Endochondral ossification.
    B) Intramenbranous ossification.
    C) Metaphysical ossification.
    D) Oscar's fictional vacation.
    E) Both A & B.
    Answer: A

45. Which is not part of the axial skeleton?
    A) facial skull bones
    B) vertebrae
    C) ribs
    D) talus
    E) cranial skull bones
    Answer: D

46. A round shaped bone composed of spongy bone which is covered with compact bone would be classified as a(n) ___ bone.
    A) flat.
    B) irregular.
    C) short.
    D) long.
    E) oval.
    Answer: C

47. The type of cartilage found in a cartilaginous joint
    A) hyaline cartilage
    B) fibrocartilage
    C) elastic cartilage
    D) dense connective tissue – it is every where else.
    E) irregular connective tissue.
    Answer: B

True-False

Write T if the statement is true and F if the statement is false.

1. The brain is protected by cranial bones.
   Answer: True

2. Vertebrae are also called Wormian bones.
Answer: False

3. The medullary cavity contains red bone marrow.
Answer: False

4. The endosteum is the lining of the medullary cavity.
Answer: True

5. Osteoblasts are mature bone cells.
Answer: False

6. Canaliculi are characteristic of spongy bone.
Answer: False

7. Bone formation is called ossification.
Answer: True

8. The replacement of cartilage by bone is called intramembranous ossification.
Answer: False

9. The epiphyseal plate allows the diaphysis of the bone to increase in diameter.
Answer: False

10. Bone continuously replaces itself throughout adult life.
Answer: True

11. The decrease of estrogen levels in females after the age of 40 increases the loss of calcium from bones.
Answer: True

12. The sphenoid bone is one of the facial bones.
Answer: False

13. The vomer is a triangular bone, which forms part of the nasal septum.
Answer: True

14. The openings between the vertebrae form the vertebral canal.
Answer: False

15. The shoulder blades are also called clavicles.
Answer: False

16. Early in life, osteoblasts work faster than osteoclasts so bone is building up. At "midlife" they somewhat equal out, but later on, the osteoclasts work faster than the osteoblasts.
Answer: True

17. In spongy bone, the osteons are irregularly arranged but in compact bones they are parallel to each other making compact bone heavier than spongy. Both bone types are strong.
Answer: True

18. Synovial fluid moistens and lubricates the cartilaginous surfaces within a synovial joint.
Answer: True

19. Red marrow it hematopoietic (blood forming).
Answer: True

20. A condition resulting from the effects of uneven roads on the patella is called flatfoot.
Answer: False

Short Answer

Write the word or phrase that best completes each statement or answers the question.

1. Hemopoiesis occurs in _____ _____ _____.
Answer: red bone marrow

2. The distal and proximal ends of long bones are called ____________.
Answer: epiphyses

3. The tough, white, fibrous membrane around the surface of a bone is called ____.
Answer: periosteum

4. Trabeculae are found in ____ bone.
Answer: spongy

5. The formation of bone within fibrous connective tissue membranes is referred to as _____.
Answer: intramembranous ossification

6. The ongoing replacement of old bone tissue by new bone tissue is called bone __________.
Answer: remodeling

7. The ribs belong to the ____ skeleton.
Answer: axial

8. The minerals responsible for the hardening of bone matrix are _____ and ______.
Answer: calcium and phosphorus

9. Elderly people can strengthen their bones by _______-_______ exercise.
Answer: weight-bearing

10. An immovable joint between the bones of the skull is called a ____.
Answer: suture

11. The suture between the frontal bone and the two parietal bones is the ______ suture.
Answer: coronal

12. The bony sockets into which teeth are set are called ____.
Answer: alveoli

13. The bone marking used as a landmark by dentists to inject anesthetics is the ____.
Answer: mental foramen

14. The smallest facial bones are the ____ bones.
Answer: lacrimal

15. The correct anatomical term for "soft spots" on infant skulls is ____.
Answer: fontanels

16. The structures found between adjacent vertebra are ____.
Answer: intervertebral discs

17. The first cervical vertebra is called ____.
    Answer: atlas

18. The medial bone of the forearm is the _____.
    Answer: ulna

19. The bones of the fingers are called the ____.
    Answer: phalanges

20. The hipbones are united to each other by a joint called ____.
    Answer: pubic symphysis

21. The correct anatomical term for kneecap is ____.
    Answer: patella

22. The collective term for the seven bones of the ankle is ____.
    Answer: tarsus

23. The disorder characterized by decrease in bone mass is called ____.
    Answer: osteoporosis

24. A lateral bending of the vertebral column is called ____.
    Answer: scoliosis

25. A deformity of the great toes is called ____.
    Answer: bunion

Matching

Choose the item from Column 2 that best matches each item in Column 1.

1. Column 1: A Vitamin D deficiency in children in which the body does not absorb calcium and phosphorus.
   Column 2: Rickets

2. Column 1: An irregular thickening and softening of the bones.
   Column 2: Paget's disease

3. Column 1: A lateral bending of the vertebral column, usually the thoracic spine.
   Column 2: Scoliosis

4. Column 1: A deformity of the great toe.
   Column 2: Bunion

5. Column 1: A disorder characterized by decreased bone mass.
   Column 2: Osteoporosis

6. Column 1: Mastoid process
   Column 2: Temporal bones

7. Column 1: Carotid foramen
   Column 2: Temporal bones

8. Column 1: Sella turcica
   Column 2: Sphenoid bone

9. Column 1: Optic foramen
   Column 2: Sphenoid bone

10. Column 1: Condylar process
    Column 2: Mandible

11. Column 1: Spinous process
    Column 2: Vertebrum

12. Column 1: Acromion
    Column 2: Scapula

13. Column 1: Coronoid fossa
    Column 2: Humerus

14. Column 1: Acetabulum
    Column 2: Hipbone

15. Column 1: Foramen magnum
    Column 2: Occipital bone

Essay

Write your answer in the space provided or on a separate sheet of paper.

1. Explain the difference between true ribs, false ribs, and floating ribs.
   Answer: True ribs have a direct anterior attachment to the sternum via the costal cartilage. The costal cartilage of false ribs does not directly attach to the sternum, and floating ribs have no anterior attachment to the sternum at all.

2. Name the portions of the sternum.
   Answer: Manubrium, body, and xiphoid process

3. List the cranial bones.
   Answer: frontal bone, parietal bones (2), temporal bones (2), occipital bone, sphenoid bone, ethmoid bone

4. Name the eight carpal bones.
   Answer: scaphoid (navicular), lunate, triquetrum, pisiform, trapezium, trapezoid, capitate, hamate

CHAPTER 7 Joints

Multiple-Choice

Choose the one alternative that best completes the statement or answers the question.

1. The scientific study of joints is called
   A) anthropology.
   B) archeology.
   C) arthrology.
   D) orthoscopy.
   E) rheumatology.
   Answer: C

2. All of the following are structural classifications of joints EXCEPT
   A) synarthrosis.
   B) fibrous.
   C) cartilaginous.
   D) synovial.
   E) All of the above are correct.
   Answer: A

3. Slightly movable joints are called
   A) synovial joints.
   B) amphiarthorses.
   C) synarthrosis.
   D) sutures.
   E) diarthroses.
   Answer: B

4. Which of the following joints is classified as a synarthrotic joint?
   A) sydesmoses
   B) diarthrosis
   C) symphysis
   D) gomphosis
   E) amphiarthrosis
   Answer: D

5. The articulation between the root of the teeth and the alveoli is an example of a
   A) synchondrosis.
   B) diarthrosis.
   C) amphiarthrosis.
   D) suture.
   E) gomphosis.
   Answer: E

6. A joint in which the connecting material is a broad, flat disc of fibrocartilage is a
   A) symphysis.
   B) syndemosis.
   C) synchondrosis.
   D) gomphosis.
   E) suture.
   Answer: A

7. Which of the following is a cartilaginous joint?
   A) diarthrosis
   B) symphyses
   C) syndesmosis
   D) suture
   E) gomphosis
   Answer: B

8. Freely moveable joints are called
   A) synarthroses.
   B) diarthroses.
   C) symphysis.
   D) amphiarthroses.
   E) gomphosis.
   Answer: B

9. A cavity between joints is characteristic for
   A) synovial joints.
   B) symphysis.
   C) fibrous joints.
   D) amphiathrosis.
   E) None of the above.
   Answer: A

10. The inner layer of the articular capsule is a
    A) fibrous membrane.
    B) ligament.
    C) articular membrane.
    D) synovial membrane.
    E) meniscus.
    Answer: D

11. All of the following are functions of the synovial fluid EXCEPT
    A) lubrication.
    B) friction reduction.
    C) protein production.
    D) removal of wastes.
    E) supply of nutrients.
    Answer: C

12. A tearing of the menisci of the knee is commonly called a torn
    A) ligament.
    B) bursa.
    C) cartilage.
    D) tendon.
    E) muscle.
    Answer: C

13. Sac-like structures that reduce friction between moving parts are called
    A) tendons.
    B) ligaments.
    C) menisci.
    D) accessory ligaments.
    E) bursae.
    Answer: E

14. A gliding movement is a movement of a(n) type of
    A) fibrous joint.
    B) cartilaginous joint.
    C) synovial joint.
    D) All of the above.
    E) None of the above.
    Answer: C

15. The joints between carpals are examples of a
    A) ball-and-socket joint.
    B) hinge joint.
    C) pivot joint.
    D) gliding joint.
    E) saddle joint.
    Answer: D

16. Which of the following is/are synovial joints?
    A) planar joints
    B) hinge joints
    C) condyloid joints
    D) none of the above
    E) all of the above
    Answer: E

17. A movement which decreases the angle between articulating bones is called a(n)
    A) flexion.
    B) extension.
    C) circumduction.
    D) adduction.
    E) abduction.
    Answer: A

18. The joint between the trochlea of the humerus and the trochlear notch of the ulna at the elbow is an example of a
    A) saddle joint.
    B) pivot joint.
    C) hinge joint.
    D) condyloid joint.
    E) gliding joint.
    Answer: C

19. Extension is a movement at a
    A) gliding joint.
    B) hinge joint.
    C) pivot joint.
    D) ball-and-socket joint.
    E) saddle joint.
    Answer: B

20. The primary movement of a pivot joint is
    A) rotation.
    B) abduction.
    C) adduction.
    D) circumduction.
    E) hyperextension.
    Answer: A

21. Up-and-down and side-to-side movements are performed by a
   A) saddle joint.
   B) pivot joint.
   C) hinge joint.
   D) condyloid joint.
   E) A or D.
   Answer: E

22. The shoulder joint is an example of a
   A) saddle joint.
   B) pivot joint.
   C) ball-and-socket joint.
   D) hinge joint.
   E) gliding joint.
   Answer: C

23. The lateral movement of the arms away from the body is called
   A) elevation.
   B) abduction.
   C) adduction.
   D) flexion.
   E) pronation.
   Answer: B

24. Which of the following is NOT considered a special movement only occurring at particular joints?
   A) elevation
   B) protraction
   C) circumduction
   D) supination
   E) dorsiflexion
   Answer: C

25. Hyperextension represents
   A) the movement of a bone away from the midline.
   B) the movement of the distal end of a part of the body in a circle.
   C) the movement of a part of the body forward.
   D) an extension beyond the anatomical position.
   E) movement of a protracted part of the body back to the anatomical position.
   Answer: D

26. The shrugging of the shoulders to elevate the scapula is an example of
   A) abduction.
   B) depression.
   C) inversion.
   D) circumduction.
   E) elevation.
   Answer: E

27. An autoimmune disease that attacks the joints is
   A) rheumatoid arthritis.
   B) Lyme disease.
   C) osteoarthritis.
   D) arthralgia.
   E) multiple sclerosis.
   Answer: A

28. The field of medicine devoted to the disease of the joints is
    A) oncology.
    B) arthroscopy.
    C) dermatology.
    D) radiology.
    E) rheumatology.
    Answer: E

29. The arcuate popliteal ligament is found in the
    A) sole of the foot.
    B) knee.
    C) elbow.
    D) hip.
    E) shoulder.
    Answer: B

30. An inflammation of the joints is called
    A) bursitis.
    B) tendonitis.
    C) arthritis.
    D) perichondritis.
    E) chondritis.
    Answer: C

31. Lyme disease is caused by
    A) fever.
    B) a virus.
    C) ticks.
    D) a bacterium.
    E) All of the above.
    Answer: D

32. What type of cartilage is articular cartilage?
    A) fibrous
    B) reticular
    C) hyaline
    D) elastic
    E) collagen
    Answer: C

33. The outer layer of the articular capsule consists of
    A) dense irregular connective tissue.
    B) dense regular connective tissue.
    C) cartilage.
    D) synovial membranes.
    E) loose connective tissue.
    Answer: A

34. A decrease of the angles between bones is achieved by
    A) flexion.
    B) extension.
    C) adduction.
    D) abduction.
    E) pronation.
    Answer: A

35. A forward movement of the mandible would be called
    A) inversion.
    B) protraction.
    C) flexion.
    D) retraction.
    E) supination.
    Answer: B

36. Bending the foot upward involves
    A) elevation.
    B) planar flexion.
    C) dorsiflexion.
    D) inversion.
    E) supination.
    Answer: C

37. All of the following are part of the knee joint EXCEPT
    A) medial meniscus.
    B) patellar ligament.
    C) articular capsule.
    D) bursae.
    E) transverse humeral ligament.
    Answer: E

38. A joint that is eventually replaced by bone is a
    A) synchondrosis.
    B) sydesmoisis.
    C) symphysis.
    D) diarthrosis.
    E) suture.
    Answer: A

39. The distal articulation of the tibia and fibula is a
    A) diarthrosis.
    B) sydesmoisis.
    C) synchondrosis.
    D) symphysis.
    E) synarthrosis.
    Answer: B

40. The displacement of a bone from a joint with the tearing of ligaments, tendons, and articular capsules is called a(n)
    A) synovitis.
    B) chondritis.
    C) arthralgia.
    D) dislocation.
    E) sprain.
    Answer: D

41. Bones in close contact with each other form which type of joint?
    A) synovial.
    B) fibrous.
    C) cartilaginous.
    D) cannibus.
    E) diatrthrotic.
    Answer: C

42. Abduction is when a body part is moved
    A) around an axis.
    B) to decrease the angle between that part and another.
    C) away from the body.
    D) in all planes but rotation.
    E) towards the body.
    Answer: C

43. Flexion is when a body part is moved
    A) around an axis.
    B) to decrease the angle between that part and another.
    C) away from the body.
    D) in all planes but rotation.
    E) towards the body.
    Answer: B

44. Mostly immovable articulations such as the suture between the parts of the skull are classified as
    A) dysfunction.
    B) synovial.
    C) fibrous.
    D) cartilaginous.
    E) illegal.
    Answer: C

45. The type of cartilage found in a cartilaginous joint is
    A) hyaline cartilage.
    B) fibrocartilage.
    C) elastic cartilage.
    D) dense connective tissue (it is everywhere else!).
    E. dense irregular connective tissue.
    Answer: B

46. Hinge joints are capable of
    A) elevation and flexion.
    B) adduction and abduction.
    C) depression and elevation.
    D) flexion and extension.
    E) all of the above.
    Answer: D

47. Which of the following is NOT an example of a hinge joint?
    A) ulna and humerus
    B) atlas and axis
    C) skull and atlas
    D) knee (femur and tibia)
    E. middle and distal phalange
    Answer: B.

48. Extension is when a body part is moved
    A) toward the body.
    B) to increase the angle between that part and another.
    C) away from the body.
    D) in all planes including rotation.
    E) to decrease the angle between that part and another.
    Answer: B

49. Freely moveable joints are lubricated by
    A) synovial fluid.
    B) mucous membranes.
    C) serous fluid.
    D) lymph.
    E) plasma.
    Answer: A

True-False

Write T if the statement is true and F if the statement is false.

1. The contact between teeth and bone is an articulation.
   Answer: True

2. A suture is a fibrous joint.
   Answer: True

3. Synarthroses are slightly moveable joints.
   Answer: False

4. Pads of fibrocartilage inside the knee joint are called articular discs.
   Answer: True

5. The epiphyseal plate is an example of a gomphosis.
   Answer: False

6. Ligaments are present in all types of articulations.
   Answer: True

7. Abduction refers to movement away from the midline of the body.
   Answer: True

8. The posterior cruciate ligament extends posteriorly from the tibia to the femur.
   Answer: False

9. Arthritis is a form of rheumatism.
   Answer: True

10. A strain is more serious than a sprain.
    Answer: False

11. Arthralgia is the displacement of a bone from a joint.
    Answer: False

12. The anterior cruciate ligament belongs to the shoulder joint.
    Answer: False

13. Pronation is a movement of the forearm in which the palm is turned backward or downward.
    Answer: True

14. Syndesmoses are slightly moveable joints.
    Answer: True

15. The cartilage in a cartilaginous joint is elastic cartilage to ensure free movement.
    Answer: False

16. Synovial fluid moistens and lubricates the cartilaginous surfaces in a diathrotic joint.
    Answer: True

17. The movement of a part away from the midline of the body (first part of jumping jacks!) is called abduction.
    Answer: True

18. The difference between rheumatoid arthritis and osteoarthritis is that osteoarthritis is an autoimmune disease.
    Answer: False

19. Lyme disease is named for the bacterium that causes it. *Lymerius cocci.*
    Answer: False

Short Answer

Write the word or phrase that best completes each statement or answers the question.

1. The field of medicine dealing with joint diseases is _____________.
   Answer: rheumatology

2. Another name for *joint* is _________________.
   Answer: articulation

3. The study of the movement of the human body is called __________.
   Answer: kinesiology

4. The space between articulating bones is called ____.
   Answer: synovial cavity

5. The displacement of the articulating bones from their normal position is called _________.
   Answer: dislocation

6. The three types of fibrous joints are sutures, syndesmoses, and _______________.
   Answer: gomphoses

7. A tearing of menisci in the knee is commonly called ______ __________.
   Answer: torn cartilage

8. A freely movable joint is classified as a(n) ____.
   Answer: diarthrosis

9. The fibrous joints between the bones of the skull are called ____.
   Answer: sutures

10. The epiphyseal plate is a type of immovable joint (synarthrosis) called ____.
    Answer: synchondrosis

11. The distal articulation of the tibia and fibula is a slightly moveable joint (amphiarthrosis) of the type ____.
    Answer: syndesmosis

12. The capsule that surrounds a synovial cavity is called ____.
    Answer: articular capsule

13. The synovial membrane secretes ____.
Answer: synovial fluid

14. An inflammation of the bursa is called ____.
Answer: bursitis

15. The type of freely movable joints in the ankle are ____ joints.
Answer: hinge

16. A movement toward the midline of the body is called a(n) ____.
Answer: adduction

17. An upward movement of a part of the body is a(n) ____.
Answer: elevation

18. The lateral meniscus belongs to the ____ joint.
Answer: knee

19. The displacement of a bone from a joint with the tearing of ligaments, tendons, and articular capsule is called a ____.
Answer: dislocation or luxation

20. The inflammation of cartilage is ____.
Answer: chondritis

21. The point of contact between two or more bones is called a(n) ____.
Answer: articulation

22. Structurally, joints are classified as synovial, fibrous, and ______.
Answer: cartilaginous

23. Functionally, joints are classified as synarthroses, diarthroses, and ____.
Answer: amphiarthroses

24. The specific type of synovial joints between the radius and the ulna is a ____ joint.
Answer: pivot

25. The specific classification given the synovial joint of the shoulder, is an example of a(n) _____ joint.
Answer: ball-and-socket

Matching

Choose the item in Column 2 that best matches each item in Column 1.

1. Column 1: An immovable joint
Column 2: synarthrosis

2. Column 1: A freely movable joint
Column 2: diarthrosis

3. Column 1: A fibrous joint found between the bones of the skull.
Column 2: suture

4. Column 1: A joint that permits movements in three planes.
Column 2: ball-and-socket joint

5. Column 1: A joint that is capable of hyperextension.
   Column 2: hinge joint

6. Column 1: A joint between the sternum and the clavicle.
   Column 2: gliding joint

7. Column 1: A joint that provides sliding or twisting movement.
   Column 2: gliding joint

Essay

Write your answer in the space provided or on a separate sheet of paper.

1. Classify the joints according to function and indicate the degree of movement each permits.
   Answer: Synarthroses are immovable joints, amphiarthroses are slightly movable joints, and Diarthroses are freely moveable joints.

2. Name the different types of diarthroses.
   Answer: gliding joints, hinge joints, pivot joints, condyloid joints, saddle joints, ball-and-socket joints

3. Name and describe the six special movements that relax the foot and hand.
   Answer: Inversion is the movement of the sole inward, eversion is the movement of the sole outward, dorsiflexion involves bending the foot downward, plantar flexion involves pointing the toes, and supination and pronation are movements of the forearm.

CHAPTER 8 The Muscular System

Multiple-Choice

Choose the one alternative that best completes the statement or answers the question.

1. Which of the following is NOT a function of muscle tissue?
   A) Heat production
   B) Regulation of organ volume
   C) Vitamin D production
   D) Body movements
   E) Body position
   Answer: C

2. The structure made of dense irregular connective tissue that holds muscles together and separates them into functional groups is
   A) epimysium.
   B) endomysium.
   C) perimysium.
   D) superficial fascia.
   E) deep fascia.
   Answer: E

3. Bundles of muscle fibers are covered by
   A) superficial fascia.
   B) deep fascia.
   C) perimysium.
   D) epimysium.
   E) endomysium.
   Answer: C

4. The portion of the sarcomere mostly composed of thick filaments is the
   A) H zone.
   B) Z discs.
   C) I band.
   D) A band.
   E) sarcomere.
   Answer: D

5. The sarcomere is the area between two
   A) Z discs.
   B) I bands.
   C) A bands.
   D) H zones.
   E) B lines.
   Answer: A

6. The pigment in the muscle fibers that stores oxygen is
   A) hemoglobin.
   B) myoglobin.
   C) melanin.
   D) melanin.
   E) glycogen.
   Answer: B

7. The neurotransmitter at the neuromuscular junction is
   A) norepinephrine.
   B) acetylcholine.
   C) adrenaline.
   D) calcium.
   E) glycine.
   Answer: B

8. Tunnel-like extensions of the sarcolemma into the muscle fiber are
   A) sarcomeres.
   B) myofibrils.
   C) transverse tubules.
   D) tropomyosin.
   E) myosin cross bridges.
   Answer: C

9. The structure storing calcium in the skeletal muscle fibers at rest is the
   A) endoplasmic reticulum.
   B) nucleus.
   C) Golgi apparatus.
   D) sarcoplasmic reticulum.
   E) T–Tubule.
   Answer: D

10. All of the following are proteins of myofilaments EXCEPT
    A) actin.
    B) myosin.
    C) troponin.
    D) tropomyosin.
    E) elastin.
    Answer: E

11. The space between an axon terminal and the sarcolemma is the
    A) neuromuscular junction.
    B) synaptic cleft.
    C) synaptic vesicle.
    D) motor end plate.
    E) motor unit.
    Answer: B

12. Which of the following substances can block the release of acetylcholine at the neuromuscular junction?
    A) acetylcholinesterase
    B) curare
    C) botulinum toxin
    D) anticholinesterases
    E) cocaine
    Answer: C

13. Which of the following substances are absolutely necessary for muscle contraction?
    A) potassium
    B) ATP and calcium
    C) ATP and sodium
    D) ADP and phosphate
    E) potassium and ATP
    Answer: B

14. Acetylcholine in the synaptic cleft is broken down by
    A) anticholinesterase.
    B) pseudocholinesterese.
    C) acetylcholinesterase.
    D) cholinesterase.
    E) None of the above.
    Answer: C

15. Muscle tone is due to
    A) the contraction of most muscle fibers in a whole muscle.
    B) a small number of motor units activated.
    C) sustained muscle contraction.
    D) None of the above.
    E) All of the above.
    Answer: B

16. Muscle fatigue can be caused by
    A) increased ADP.
    B) insufficient oxygen.
    C) depletion of creatine phosphate.
    D) A and B.
    E) All of the above.
    Answer: E

17. Most of the lactic acid produced by anaerobic respiration in muscle fibers is reconverted to glucose in the
    A) muscle fibers.
    B) liver.
    C) pancreas.
    D) kidney.
    E) blood.
    Answer: B

18. A sustained contraction of a muscle is called
    A) tetanus.
    B) summation.
    C) latent period
    D) twitch.
    E) refractory period.
    Answer: A

19. An increase in the diameter of muscle fibers is called
    A) muscular atrophy.
    B) muscular hypertrophy.
    C) hypertonia.
    D) hypotonia.
    E) Both A and D are correct.
    Answer: B

20. Which of the following is/are striated and involuntary?
    A) skeletal muscle
    B) cardiac muscle
    C) smooth muscle
    D) both a and b are correct
    E) both a and c are correct
    Answer: B

21. Intercalated discs are characteristic structures in
    A) cardiac muscle.
    B) smooth muscle.
    C) skeletal muscle.
    D) any striated muscle.
    E) striated and smooth muscle.
    Answer A

22. Intermediate filaments are structures of
    A) cardiac muscle.
    B) smooth muscle.
    C) skeletal muscle.
    D) striated muscle.
    E) skeletal and smooth muscle.
    Answer: B

23. Inflammation of muscle fibers is called
    A) myalgia.
    B) myoma.
    C) myositis.
    D) paralysis.
    E) myospasm.
    Answer: C

24. The ability of muscle fibers to shorten is referred to as
    A) excitability.
    B) contractility.
    C) extensibility.
    D) elasticity.
    E) recoiling.
    Answer: B

25. The immediate direct source of energy for muscle contraction is
    A) creatine phosphate.
    B) myoglobin.
    C) ATP.
    D) ADP.
    E) glycogen.
    Answer: C

26. All of the following are considered a kind of contraction EXCEPT
    A) tetanus.
    B) twitch.
    C) isotonic.
    D) isometric.
    E) All of the above are contractions.
    Answer: E

27. The attachment of a muscle to the stationary bone is called
    A) insertion.
    B) fixation.
    C) head.
    D) origin.
    E) ligation.
    Answer: D

28. Which of the following is NOT a characteristic used to name skeletal muscles?
A) direction of fibers
B) thickness of fibers
C) location
D) size
E) number of origins
Answer: B

29. A muscle that has three origins is called a
A) bicep.
B) tricep.
C) quadricep.
D) deltoid.
E) trapezius.
Answer: B

30. A muscle that produces an upward movement is a
A) pronator.
B) tensor.
C) flexor.
D) depressor.
E) levator.
Answer: E

31. A muscle that decreases the size of an opening is a
A) rotator.
B) tensor.
C) pronator.
D) sphincter.
E) depressor.
Answer: D

32. Which of the following is a muscle of the facial expression?
A) eternal oblique
B) rhomboideus
C) teres major
D) platysma
E) pronator teres
Answer: D

33. The muscle that produces the action of sucking is the
A) buccinator.
B) zygomaticus.
C) frontalis.
D) epicranius.
E) occipitalis.
Answer: A

34. Which of the following is a muscle that moves the mandible?
A) orbicularis occuli
B) masseter
C) orbicularis oris
D) epicranius
E) frontalis
Answer: B

35. All of the following are muscles that move the eyeball EXCEPT
    A) superior rectus.
    B) lateral rectus.
    C) inferior oblique.
    D) internal oblique.
    E) superior oblique.
    Answer: D

36. Which of the following is a muscle that moves the pectoral girdle?
    A) trapezius
    B) pectoralis
    C) latissimus dorsi
    D) gracillus
    E) sartorius
    Answer: A

37. All of the following are muscles that move the humerus EXCEPT
    A) latissimus dorsi.
    B) infraspinatus.
    C) teres major.
    D) deltoid.
    E) pronator teres.
    Answer: E

38. Which of the following flexes and abducts the wrist?
    A) flexor digitorum profundus
    B) flexor digitorum superficialis
    C) flexor carpi radialis
    D) flexor carpi ulnaris
    E) palmaris longus
    Answer: C

39. The sternocleidomastoid is a muscle that moves the
    A) fingers.
    B) mandible.
    C) maxilla.
    D) humerus.
    E) vertebral column.
    Answer: E

40. All of the following are muscles of the lower limb EXCEPT
    A) extensor digitorum.
    B) adductor magnus.
    C) adductor longus.
    D) quadriceps femoris.
    E) vastus medialis.
    Answer: A

41. Which of the following is a muscle that moves the foot?
    A) peroneus magnus
    B) semitendinosus
    C) sartorius
    D) biceps femoris
    E) gracilis
    Answer: A

42. Which of the following muscles adducts the thigh and flexes the leg?
A) rectus femoris
B) gracilis
C) tibialis anterior
D) gastronemicus
E) flexor digitorum longus
Answer: B

43. The infraspinatus muscle is a muscle that moves the
A) upper leg.
B) lower leg.
C) upper arm.
D) lower arm.
E) carpals.
Answer: C

44. Pain in or associated with muscles is called
A) hypertonia.
B) proprioception.
C) myotonia.
D) myalgia.
E) myositis.
Answer: D

45. When a skeletal muscle fiber normally contracts, it
A) contracts completely or not at all.
B) contracts as a twitch.
C) needs a subthreshold stimulus.
D) conducts the impulse from cell to cell.
E) all of the above are true.
Answer: A

46. A motor unit is all the:
A) muscle fascicles involved in a muscle contraction.
B) muscle fibers controlled by a single motor neuron.
C) myofibrils that contract in a single nerve impulse.
D) muscles involved in a single movement.
E) none of the above.
Answer: B

47. The difference between picking up a pencil and 30 pounds is
A) the number of antagonists involved.
B) the size of the motor units involved.
C) the frequency of nerve impulses.
D) the number of motor units involved.
E) B & D.
Answer: E

48. The number of muscle fibers in a motor unit is
A) greater where precise control is needed.
B) lesser where precise control is needed.
C) lesser in muscles for posture.
D) greater in muscles moving the fingers.
E) Both B & D.
Answer: B

49. Which is true of smooth muscle?
    A) it is not under voluntary control
    B) it contracts faster than skeletal muscle
    C) it fatigues faster than skeletal muscle
    D) each muscle fiber must be innervated
    E) all are true.
    Answer: A

50. The amount of oxygen needed to support the conversion of lactic acid to glucose is called the
    A) muscle fatigue.
    B) oxygen debt.
    C) anaerobic conversion.
    D) aerobic conversion.
    E) summation.
    Answer: B

51. The cell membrane of a muscle cell is named
    A) sarcoplasmic reticulum,
    B) sarcolemma.
    C) sarcoplasm..
    D) George.
    E) sarcasm.
    Answer: B

52. Creatine Phosphate converts
    A) ATP to ADP.
    B) ADP to ATP.
    C) deoxymyoglobin to myoglobin.
    D) lactic acid to pyruvic acid.
    E) pyruvic acid to lactic acid.
    Answer: B

53. Smooth muscle is
    A) striated, multinucleated and voluntary.
    B) non-striated, single nucleated and involuntary.
    C) striated, single nucleated and involuntary.
    D) non-striated, multinucleated and involuntary.
    E) striated, multinucleated and involuntary.
    Answer: B

54. Cardiac muscle is
    A) striated, multinucleated and voluntary.
    B) non-striated, single nucleated and involuntary.
    C) striated, single nucleated and involuntary.
    D) non-striated, multinucleated and involuntary.
    E) striated, multinucleated and involuntary.
    Answer: C

True-False

Write T if the statement is true and F if the statement is false.

1. Skeletal muscle tissue is striated and voluntary.
   Answer: True

2. One byproduct of skeletal muscle contraction is heat.
Answer: True

3. The myofibrils of skeletal muscle fibers are composed of three types of myofilaments.
Answer: False

4. The H zone is located in the center of the I band.
Answer: False

5. During muscle contraction, the myosin heads pull on the actin filaments, causing the actin to slide toward the center of the sarcomere.
Answer: True

6. The ends of axon terminals enlarge into swellings called motor end plates.
Answer: False

7. The power stroke in muscle contraction is the force causing the actin filaments to slide between the myosin filaments.
Answer: True

8. The detachment of myosin heads from actin requires ATP.
Answer: True

9. Any stimulus from a neuron can initiate a muscle contraction.
Answer: False

10. Muscle tissue plays a major role in maintenance of homeostasis.
Answer: True

11. Skeletal muscle has a long refractory period.
Answer: False

12. When a muscle shortens and pulls on another structure, it is undergoing an isotonic contraction.
Answer: True

13. Muscle tone requires the contraction of an entire muscle.
Answer: False

14. There are two types of smooth muscle tissue, visceral and multiunit.
Answer: True

15. Smooth muscle is under voluntary control.
Answer: False

16. A muscle fiber can partially contract.
Answer: False

17. Each skeletal muscle fiber has its own nerve fiber.
Answer: False

18. A muscle (such as the lovely gluteus maximus) is made of bundles of fascicles which are bundles of muscle fibers which are bundles of myofibrils which are arrangements of myofilaments
Answer: True

19. Some muscles, called antagonists, contract to work together. They are aided by prime movers. While this happens, opposite muscles called synergists relax to support the bone.
Answer: False

20. Both smooth and cardiac muscle are involuntary and contain actin and myosin filaments. However, the actin and myosin in the smooth muscle are arranged in stripes (similar to skeletal muscle) but they are irregularly arranged in the cardiac muscle.
    Answer: False

21. The muscle attachment to the bone that moves is called the origin.
    Answer: False

Short Answer

Write the word or phrase that best completes each statement or answers the question.

1. The sheet of fibrous connective tissue around muscles is __________.
   Answer: fascia

2. The electrical current that travels over the sarcolemma and along the transverse tubules is called ____.
   Answer: muscle action potential

3. The cytoplasm of a muscle fiber is called ____.
   Answer: sarcoplasm

4. The basic, functional units of skeletal muscle fibers are ____.
   Answer: sarcomeres

5. During muscle contraction filaments ____, they do not shorten.
   Answer: slide

6. A motor neuron and the muscle fibers it stimulates is called a ____ ____.
   Answer: motor unit

7. Synaptic vesicles are filled with chemicals called ____.
   Answer: neurotransmitter

8. The ion absolutely essential for the sliding of the filaments in muscle contraction is ____.
   Answer: calcium

9. A poison used by South American Indians to cause muscle paralysis is __________.
   Answer: curare

10. ____ is a high energy molecule in skeletal muscle fibers, which can be used to produce ATP as needed.
    Answer: creatine phosphate

11. In order for a muscle fiber to contract, a ____ stimulus has to be applied.
    Answer: threshold

12. In response to a stimulus, muscle fibers either contract or do not contract, which is in accordance with the ____ principle.
    Answer: all-or-none

13. The inability of a muscle to maintain its strength is called ____.
    Answer: (muscle) fatigue

14. The period of time between the application of a stimulus and the beginning of muscle contraction is called _____.
    Answer: latent period

15. A contraction where the tension of the muscle increases, but the shortening is minimal, is called a(n) _____ contraction.
Answer: isometric

16. The intermediate filaments in smooth muscle fibers stretch between structures called ____ ____.
Answer: dense bodies

17. A muscle that causes a desired action is referred to as the _____ ____.
Answer: prime mover (agonist)

18. Derivatives of testosterone that are illegally used to build up muscle proteins are called ____ steroids.
Answer: anabolic

19. The muscle that moves a bone around its longitudinal axis is called a ____.
Answer: rotator

20. The maximum ability to move bones about the joint through an arc of a circle is a joint's ________ _____ ___________.
Answer: range of motion

21. The autoimmune disorder involving the neuromuscular junction is ____ ____.
Answer: myasthenia gravis

22. At death, when ATP is not available for the detachment of the myosin head from the actin, ____ ____ occurs.
Answer: rigor mortis

23. Epimysium, perimysium, and endomysium extend beyond the muscle as a ____.
Answer: tendon

24. A better term for oxygen debt is _____ ______ ______.
Answer: recovery oxygen uptake

25. The process in which the number of contracting motor units is increased is called _____ _____ ______.
Answer: motor unit recruitment

Matching

Choose the item from Column 2 that best matches each item in Column 1.

1. Column 1: Nonstriated, involuntary muscle
   Column 2: smooth muscle

2. Column 1: Striated involuntary muscle
   Column 2: cardiac muscle

3. Column 1: Consists of mostly thick filaments
   Column 2: A band

4. Column 1: Consists of mostly thin filaments
   Column 2: I band

5. Column 1: It is located in the center of each A band
   Column 2: H zone

6. Column 1: A type of neuron that stimulates muscle tissue
   Column 2: motor neuron

7. Column 1: The region of the sarcolemma near the axon terminal
   Column 2: motor end plate

8. Column 1: A process that does not require oxygen
   Column 2: anaerobic respiration

9. Column 1: A brief contraction of muscle fibers
   Column 2: Twitch

10. Column 1: Continued, sustained smooth contraction due to rapid stimulation
    Column 2: tetanus

Essay

Write your answer in the space provided or on a separate sheet of paper.

1. Name and describe the three types of muscle tissue.
   Answer: Skeletal muscle tissue is striated and under voluntary control. Cardiac muscle tissue is also striated but is involuntary. Cardiac muscle fibers are connected by intercalated discs. Smooth muscle fibers are nonstriated and involuntary.

2. Describe the connective tissue components of skeletal muscle tissue.
   Answer: An entire muscle is surrounded by a connective tissue covering called epimysium, the individual fascicles of a muscle are surrounded by perimysium, and individual muscle fibers are surrounded by endomysium.

3. Define muscle tone.
   Answer: A sustained partial contraction of portions of skeletal muscle results in muscle tone, which is essential for the maintenance of posture.

CHAPTER 9 Nervous Tissue

Multiple-Choice

Choose the one alternative that best completes the statement or answers the question.

1. The science that deals with the normal functioning and the disorders of the nervous system is called
   A) pathology.
   B) endocrinology.
   C) neuroscience.
   D) neurology.
   E) neuropathology.
   Answer: D

2. The nervous system and the _________ system share the greatest responsibility for maintaining homeostasis.
   A) immune
   B) endocrine
   C) cardiovascular
   D) respiratory
   E) sensory
   Answer: B

3. Neurons that conduct nerve impulses from the receptors to the central nervous system are
   A) motor neurons.
   B) efferent neurons.
   C) interneurons.
   D) sensory neurons.
   E) association neurons.
   Answer: D

4. Processes that carry impulses to another neuron are
   A) dendrites.
   B) axons.
   C) synapses.
   D) axon collaterals.
   E) myelin sheaths.
   Answer: B

5. Which of the following is an example of an effector?
   A) interneuron
   B) receptor
   C) efferent neuron
   D) nerve
   E) glands
   Answer: E

6. All of the following are part of the PNS EXCEPT
   A) neuroendocrine system.
   B) somatic nervous system.
   C) autonomic nervous system.
   D) enteric nervous system.
   E) sympathetic branch.
   Answer: A

7. Enteric motor neurons are responsible for
   A) contraction of skeletal muscle close to the GI tract.
   B) contraction of GI tract smooth muscle.
   C) contraction of smooth muscle of the blood vessels.
   D) Both A and B.
   E) Both B and C.
   Answer: B

8. The neuroglia that produce myelin in the peripheral nervous system are
   A) Schwann cells.
   B) oligodendrocytes.
   C) microglia.
   D) astrocytes.
   E) pituicytes.
   Answer: A

9. A group of nerve fibers in the central nervous system is called a(n)
   A) axon.
   B) nerve.
   C) tract.
   D) ganglion.
   E) nucleus.
   Answer: C

10. Groups of neuron cell bodies in the peripheral nervous system are called
    A) ganglia.
    B) nuclei.
    C) horns.
    D) tracts.
    E) nerves.
    Answer: A

11. A disorder of the central nervous system would involve
    A) peripheral nerves.
    B) senses.
    C) brain.
    D) spinal cord.
    E) brain and spinal cord.
    Answer: E

12. Which of the following cells does NOT functionally belong with the others?
    A) efferent neurons
    B) oligodendrocytes
    C) afferent neurons
    D) motor neuron
    E) interneuron
    Answer: B

13. The portion of the nervous system that is considered involuntary is the
    A) somatic nervous system.
    B) sensory nervous system.
    C) autonomic nervous system.
    D) motor nervous system.
    E) peripheral nervous system.
    Answer: C

14. The portion where an axon joins the cell body is the
    A) initial segment.
    B) axon hillock.
    C) myelin sheath.
    D) neurolemma.
    E) axon collateral.
    Answer: B

15. All of the following are functions of the nervous system EXCEPT
    A) senses changes.
    B) analyzes changes.
    C) stores potassium.
    D) responses to changes.
    E) integrates impulses.
    Answer: C

16. Compared to unmyleninated axons, myelinated axons
    A) conduct impulses faster.
    B) conduct impulses more often.
    C) conduct impulses slower.
    D) produce larger action potentials.
    E) produce longer lasting action potentials.
    Answer: A

17. The different charge between the outside and the inside of a neuron at rest is called
    A) action potential.
    B) synaptic potential.
    C) excitatory postsynaptic potential.
    D) resting membrane potential.
    E) equilibrium potential.
    Answer: D

18. The ability of nerve cells to respond to stimuli and to convert them into nerve impulses is called
    A) depolarization.
    B) hyperpolarization.
    C) excitability.
    D) threshold.
    E) contractility.
    Answer: C

19. The neuroglial cell involved in the blood brain barrier is the
    A) astrocyte.
    B) oligodendrocyte.
    C) microglia.
    D) satellite glia.
    E) ependymal cell.
    Answer: A

20. Saltatory conduction refers to the conduction of impulses in
    A) cardiac muscle.
    B) skeletal muscle.
    C) unmyelinated fibers.
    D) myelinated fibers.
    E) all axons.
    Answer: D

21. Depolarization of a membrane begins with the
    A) opening of chlorine channels.
    B) opening of sodium channels.
    C) opening of potassium channels.
    D) closing of potassium channels.
    E) closing of sodium channels.
    Answer: B

22. The stage in an action potential that immediately follows depolarization is
    A) superpolarization.
    B) polarization.
    C) repolarization.
    D) threshold.
    E) the resting period.
    Answer: C

23. The period of time during which the neuron cannot generate another action potential is called
    A) hyperpolarization.
    B) threshold period.
    C) repolarization.
    D) refractory period.
    E) depolarization.
    Answer: D

24. A stimulus strong enough to generate a nerve impulse is called
    A) threshold stimulus.
    B) action potential.
    C) subthreshold stimulus.
    D) conducting stimulus.
    E) Both A and D are correct.
    Answer: A

25. The speed of nerve impulse conduction is determined by all of the following EXCEPT
    A) temperature.
    B) fiber diameter.
    C) presence of myelin.
    D) absence of myelin.
    E) stimulus strength.
    Answer: E

26. The junction between two nerve cells is called
    A) neuromuscular junction.
    B) neuroglandular junction.
    C) gap junction.
    D) synapse.
    E) synaptic terminal.
    Answer: D

27. The special mode of impulse travel is called
    A) propagation.
    B) conduction.
    C) refraction.
    D) A and B.
    E) A and C.
    Answer: D

28. Neurotransmitters are stored in
   A) the synaptic cleft.
   B) synaptic vesicles.
   C) the postsynaptic neuron.
   D) Nissl bodies.
   E) smooth endoplasmic reticulum.
   Answer: B

29. Which of the following is necessary for the release of the neurotransmitter into the synaptic cleft?
   A) sodium
   B) iron
   C) calcium
   D) potassium
   E) magnesium
   Answer: C

30. What is the result of inhibitory neurotransmission on the postsynaptic membrane?
   A) a generator potential
   B) an action potential
   C) depolarization
   D) hyperpolarization
   E) excitation
   Answer: D

31. Neurotransmitters are released at the
   A) dendrite.
   B) axon collateral.
   C) axon terminal.
   D) axon hillock.
   E) cell body.
   Answer: C

32. Which of the following provides myelin for axons in the central nervous system?
   A) Schwann cells
   B) neurolemmocytes
   C) astrocytes
   D) microglia
   E) oligodendrocytes
   Answer: E

33. Processes that receive impulses and conduct them toward the cell body are
   A) dendrites.
   B) axons.
   C) axon collaterals.
   D) synaptic end bulbs.
   E) neurolemmocytes.
   Answer: A

34. Clusters of neuronal cell bodies and dendrites in the brain are called
   A) ganglia.
   B) tracts.
   C) regions.
   D) nuclei.
   E) columns.
   Answer: D

35. Bundles of unmyelinated axons, dendrites, and neuron cell bodies are called
A) nuclei.
B) white matter.
C) gray matter.
D) columns.
E) tracts.
Answer: C

36. The duration of a nerve impulse is approximately
A) one second.
B) a tenth of a second.
C) a microsecond.
D) a millisecond.
E) a fifth of a second.
Answer: D

37. Which of the following is an inhibitory neurotransmitter?
A) glutamate
B) glycine
C) dopamine
D) acetylcholine
E) serotonin
Answer: B

38. When a neuron is said to be depolarized
A) the Na+ is inside and the K+ is outside the membrane.
B) the Na+ is outside and the K+ is inside the membrane.
C) the Na+ and K+ are inside the membrane.
D) the Na+ and K+ are outside the membrane.
E) The Na+ outside equals the K+ inside.
Answer: C

39. The __ nervous system only contains all the sensory neurons and the voluntary motor neurons.
A) somatic.
B) central.
C) autonomic.
D) enteric.
E) parasympathetic.

40. Which of the following is the ion channel used to generate and conduct action potentials?
A) leakage channels.
B) voltage-gated channels.
C) gated channels.
D) television channels.
E) stoppage channels.
Answer: B

41. Neurons conducts impulses completely. This is called the
A) refractory period.
B) hyperpolarization.
C) All or none principle.
D) electrical excitability.
E) threshold.
Answer: C

42. An autoimmune disease which results in the destruction of myelin sheaths is called
    A) epilepsy
    B) muscular dystrophy.
    C) multiple sclerosis.
    D) endorphinism.
    E) None of the above.
    Answer: C

True-False

Write T if the statement is true and F if the statement is false.

1. The branch of medicine that deals with the functioning and disorders of the nervous system is called neurophysiology
   Answer: False

2. The cranial nerves belong to the peripheral nervous system.
   Answer: True

3. Ependymal cells are neuroglial cells.
   Answer: True

4. In general, neurons in adults cannot reproduce.
   Answer: True

5. Repolarization is due to the opening of sodium channels.
   Answer: False

6. If the resting membrane potential becomes more negative, the membrane is said to be depolarized.
   Answer: False

7. Another term for nerve impulse is action potential.
   Answer: True

8. Muscle action potentials are similar to nerve action potentials.
   Answer: True

9. Continuous conduction occurs along the entire membrane of myelinated axons.
   Answer: False

10. When nerve fibers are cooled, impulse conduction occurs at higher speeds.
    Answer: False

11. Some neural regeneration can occur in the peripheral nervous system.
    Answer: True

12. Subthreshold stimulation can cause action potentials.
    Answer: False

13. Neurotransmitters always cause hyperpolarization of the synaptic membrane.
    Answer: False

14. Excitatory neurotransmitters make an action potential less likely to occur.
    Answer: False

15. Whether an impulse crosses a synapse or not is going to depend on the amount and ratio of the inhibitory and excitatory neurotransmitters.
Answer: True

16. Saltatory conduction occurs only in myelinated axons when a nerve impulse jumps from one Node of Ranvier to the next.
Answer: True

17. The loss of the covering produced by the oligodendrocytes or Shwann cells is called demyelination.
Answer: True

18. Natural pain killers called endorphins and classified as neuropeptides.
Answer: True

Short Answer

Write the word or phrase that best completes each statement or answers the question.

1. The nervous system has two principal divisions, the ____ and the ____.
Answer: central nervous system, peripheral nervous system

2. Nerve cells that conduct impulses from the central nervous system to muscles and glands are ____ neurons.
Answer: motor or efferent

3. Side branches of axons are called ____.
Answer: axon collaterals

4. Sacs that store chemicals in the axon terminals area ____.
Answer: synaptic vesicles

5. A covering around axons, which is produced by neuroglia, is called the ____.
Answer: myelin sheath

6. The network of neurons in the walls of the organs of the gastrointestinal tract regulating the digestive system is the _______ __________.
Answer: enteric plexus

7. The gaps in the myelin sheath along axons are called _____.
Answer: neurofibral nodes (nodes of Ranvier)

8. The sodium-potassium pump is a(n) ____ transport mechanism.
Answer: active

9. The different charge on the inside and outside of a resting neuron is called ____.
Answer: resting membrane potential

10. The critical level of depolarization that has to be reached for an action potential to occur is the ____.
Answer: threshold

11. When the polarity of a membrane becomes more negative than the resting membrane potential, the membrane is considered to be ____.
Answer: hyperpolarized

12. The type of impulse conduction in myelinated fibers is called ____.
    Answer: salutatory

13. The neuron before a synapse is a ____ neuron, the neuron located after a synapse is a ____.
    Answer: presynaptic, postsynaptic

14. Axon terminals end in bulb-like structures called ____.
    Answer: synaptic end bulbs

15. The capacity of cells to repair or replicate themselves is called ____.
    Answer: regeneration

16. Large diameter fibers conduct impulses _____ than small diameter fibers.
    Answer: faster

17. The chemicals stored in synaptic vesicles are referred to as ____.
    Answer: neurotransmitters

18. The space between a presynaptic and postsynaptic neuron is called ____.
    Answer: synaptic cleft

19. A nerve cell transmits action potentials according to the ____ principle.
    Answer: all-or-none

20. Channels that open in response to a change in the membrane potential are _______ _______ _________.
    Answer: voltage-regulated channels

21. Synapses which conduct action potentials directly from the presynaptic to the postsynaptic cell are ____________ synapses.
    Answer: electrical

22. A progressive destruction of the myelin sheaths of neurons in the CNS is ________ _________.
    Answer: multiple sclerosis

Matching

Choose the item from Column 2 that best matches each item in Column 1.

1. Column 1: The time during which the neuron cannot generate another action potential
   Column 2: refractory period

2. Column 1: Also called a nerve impulse
   Column 2: action potential

3. Column 1: The loss of polarization
   Column 2: depolarization

4. Column 1: An increase in the negativity of the membrane potential
   Column 2: hyperpolarization

5. Column 1: The ability of a cell to respond to stimuli and to convert them into nerve impulses
   Column 2: excitability

6. Column 1: An insufficient stimulus
   Column 2: subthreshold stimulus

7. Column 1: Collection of nerve fibers in the CNS
   Column 2: tracts

8. Column 1: A collection of nerve cell bodies in the PNS
   Column 2: ganglia

9. Column 1: Produces myelin in the PNS
   Column 2: neurolemmocyte

10. Column 1: Carries nerve impulses away from the cell body
    Column 2: axon

11. Column 1: Carries impulses to the cell body
    Column 2: dendrite

Essay

Write your answer in the space provided or on a separate sheet of paper.

1. Name and describe the function of the different cytoplasmic processes in neurons.
   Answer: Axons carry impulses away from the cell body to another neuron or tissue, while dendrites carry impulses to the cell body.

2. Describe nerve impulse conduction in unmyelinated and myelinated axons.
   Answer: In unmyelinated axons, impulses travel by continuous conduction which means that the action potential in one area causes depolarization of the adjacent area. This step-by-step depolarization is slower conduction than occurs in myelinated axons. In myelinated axons, depolarization occurs at the nodes of Ranvier and impulses jump from one node to the next. This conduction is called saltatory conduction.

3. Name the six types of neuroglia and give their general function.
   Answer: Astrocytes form part of the blood brain barrier, oligodendrocytes provide the myelin in the CNS, neuroglia are phagocytotic cells in the CNS, ependymal cells line the brain ventricles, neurolemmocytes (Schwann cells) provide the myelin in the PNS, and satellite cells are supportive glia in the PNS.

CHAPTER 10 Central Nervous System, Spinal Nerves, and Cranial Nerves

Multiple-Choice

Choose the one alternative that best completes the statement or answers the question.

1. The subarachnoid space is located between the
   A) arachnoid and pia mater.
   B) arachnoid and dura mater.
   C) dura mater and pia mater.
   D) bone and pia mater.
   E) bone and arachnoid.
   Answer: A

2. The outermost of the meninges is the
   A) pia mater.
   B) Arachnoid mater
   C) dura mater.
   D) Subarachnoid space.
   E) subdural layer.
   Answer: C

3. The dura mater is composed of
   A) smooth muscle.
   B) adipose tissue.
   C) loose connective tissue.
   D) dense regular connective tissue.
   E) dense irregular connective tissue.
   Answer: E

4. The cervical enlargement contains nerves that supply the
   A) upper limbs.
   B) lower limbs.
   C) thorax.
   D) vertebral column.
   E) None of the above.
   Answer: A

5. The axillary nerve belongs to the
   A) cervical plexus.
   B) brachial plexus.
   C) intercostal nerves.
   D) lumbar plexus.
   E) sacral plexus.
   Answer: B

6. Which of the following is NOT a term used to describe a spinal cord structure?
   A) central canal
   B) anterior median fissure
   C) corpus callosum
   D) lumbar enlargement
   E) cauda equina
   Answer: C

7. The dorsal root contains
   A) sensory fibers only.
   B) motor fibers only.
   C) both sensory and motor fibers.

D) autonomic fibers only.
E) somatic fibers only.
Answer: A

8. In the reflex arc, a muscle or gland is considered to be the
A) receptor.
B) integrating center.
C) sensor.
D) motor neuron.
E) effector.
Answer: E

9. Autonomic reflexes involve
A) skeletal muscle.
B) smooth muscle.
C) cardiac muscle.
D) Both A and B.
E) Both B and C.
Answer: E

10. Which of the spinal nerves do NOT form a plexus?
A) C1-C5
B) C5-T1
C) T2-T11
D) T12-L4
E) L4-S4
Answer: C

11. Which spinal cord plexus supplies the abdominal wall?
A) cervical plexus
B) sacral plexus
C) brachial plexus
D) lumbar plexus
E) brachial and sacral plexuses
Answer: D

12. Damage to the choroid plexus would interfere with
A) myelin production.
B) formation of cerebrospinal fluid.
C) glycogen storage.
D) motor movements.
E) impulse conduction.
Answer: B

13. The portion of the brain which is continuous with the spinal cord is the
A) cerebrum.
B) pons.
C) medulla.
D) midbrain.
E) cerebellum.
Answer: C

14. The cardiovascular center is located in the
   A) hippocampus.
   B) cerebrum.
   C) diencephalon.
   D) medulla oblongata.
   E) insula.
   Answer: D

15. The thalamus is part of the
   A) brain stem.
   B) diencephalon.
   C) cerebrum.
   D) cerebellum.
   E) hypothalamus.
   Answer: B

16. The nucleus of cranial nerve number VII (facial nerve) is located in the
   A) pons.
   B) medulla.
   C) midbrain.
   D) cerebellum.
   E) cerebrum.
   Answer: A

17. The cerebral peduncles are part of the
   A) pons.
   B) medulla.
   C) midbrain.
   D) insula.
   E) hypothalamus.
   Answer: C

18. Which of the following areas deals with emotion?
   A) reticular activating system
   B) thalamus
   C) brain stem
   D) limbic system
   E) cerebrum
   Answer: D

19. Which of the areas of the cerebral cortex receives sensation from cutaneous and muscular receptors?
   A) association areas
   B) primary somatosensory area
   C) primary motor area
   D) gnostic area
   E) premotor area
   Answer: B

20. Which of the following is NOT a ventricle in an adult brain?
   A) lateral ventricle
   B) first ventricle
   C) third ventricle
   D) fourth ventricle
   E) All of the above are ventricles in the adult brain.
   Answer: B

21. The primary visual area is located in the
    A) occipital lobe.
    B) temporal lobe.
    C) frontal lobe.
    D) Broca's area.
    E) lateral lobe.
    Answer: A

22. All of the following areas are associated with memory EXCEPT
    A) parts of the limbic system.
    B) parts of the diencephalon.
    C) frontal lobe.
    D) parietal, occipital, and temporal lobes.
    E) cerebellum.
    Answer: E

23. Which of the following numbers belongs to the trochlear nerve?
    A) I
    B) III
    C) IV
    D) VII
    E) X
    Answer: C

24. Which structure does NOT belong with the others?
    A) medulla
    B) pons
    C) brain stem
    D) thalamus
    E) midbrain
    Answer: D

25. Which of the following cranial nerves is sensory only?
    A) olfactory
    B) oculomotor
    C) facial
    D) glossopharyngeal
    E) hypoglossal
    Answer: A

26. The cranial nerve that innervates the gastrointestional tract is the
    A) hypoglossal (XII).
    B) accessory (XI).
    C) vagus (X).
    D) abducens (VI).
    E) None of the cranial nerves innervate the gastrointestinal tract.
    Answer: C

27. The area of the brain that regulates posture and balance is the
    A) limbic system.
    B) cerebrum.
    C) thalamus.
    D) cerebellum.
    E) brain stem.
    Answer: D

28. In a spinal tap, cerebrospinal fluid is removed from the
    A) epidural space.

B) foramen magnum.
C) subdural space.
D) fourth ventricle.
E) subarachnoid space.
Answer: E

29. All of the following are part of the reflex arc EXCEPT
A) sensory neuron.
B) motor neuron.
C) hormones.
D) effectors.
E) nerve impulses.
Answer: C

30. The intercostal nerves innervate the
A) lower limbs.
B) muscles between the ribs.
C) muscles of the upper arm.
D) lungs.
E) shoulder.
Answer: B

31. The pyramids of the brain belong to the
A) diencephalon.
B) cerebrum.
C) cerebellum.
D) medulla oblongata.
E) hypothalamus.
Answer: D

32. The IIIrd and IVth cranial nerves originate in the
A) reticular formation.
B) thalamus.
C) midbrain.
D) pons.
E) medulla.
Answer: C

33. A shallow groove on the surface of the cortex is a
A) fissure.
B) sulcus.
C) gyrus.
D) furrow.
E) indentation.
Answer: B

34. All of the following are functions of the hypothalamus EXCEPT
A) control of the pituitary gland.
B) regulation of eating.
C) regulation of respiration.
D) regulation of body temperature.
E) satiety center.
Answer: C

35. Parkinson's disease is due to the degeneration of dopamine-producing neurons in the
A) cerebellum.
B) substantia nigra.
C) medulla.

D) hippocampus.
E) cerebral peduncles.
Answer: B

36. The term for pain relief is
A) analgesia.
B) agnosia.
C) anethesia.
D) dementia.
E) nerve block.
Answer: A

37. The primary auditory area is located in the
A) occipital lobe.
B) postcentral gyrus.
C) frontal lobe.
D) temporal lobe.
E) parietal lobe.
Answer: D

38. Which of the cranial nerves is the optic nerve?
A) I
B) II
C) III
D) VI
E) V
Answer: B

39. The meninges layer which is closest to the bone is called the
A) dura mater.
B) subarachnoid space.
C) pia mater.
D) arachnoid mater.
E) paranoid mater.
Answer: C

40. The part of the brain responsible for relaying information to and from the cerebrum and for cognition is the
A) hypothalamus.
B) pons.
C) cerebellum.
D) thalamus.
E) occipital lobe.
Answer: D

41. Which of the following is NOT a covering of the nerves?
A) epineurium.
B) endoneurium.
C) endomysium
D) perineurium.
E) All are coverings of nerves.
Answer: C

42. If someone suffered a stroke which interfered with their ability to speak, we would assume it affected the
A) temporal lobe.
B) parietal lobe.
C) association area of the frontal lobe.

D) occipital lobe.
E) motor areas of the frontal lobe.
Answer: E

43. Somatic motor neuron cell bodies are located in the:
A) lateral gray horn.
B) anterior gray horn.
C) dorsal gray horn.
D) anterior white column.
E) posterior white column.
Answer: B

44. The herpes zoster virus causes
A) meningitis.
B) neuralgia.
C) shingles.
D) agnosia.
E) encephalitis.
Answer: C

True-False

Write T if the statement is true and F if the statement is false.

1. The middle layer of the meninges is called pia mater.
Answer: False

2. The cervical enlargement contains nerves that supply the upper extremities.
Answer: True

3. Pyramidal pathways convey nerve impulses that program autonomic movements.
Answer: False

4. The dorsal root ganglion contains cell bodies of sensory neurons.
Answer: True

5. The withdrawal reflex is an autonomic reflex.
Answer: False

6. Spinal nerves are mixed nerves.
Answer: True

7. The brain contains approximately 100 million neurons.
Answer: False

8. There are four lateral ventricles in the brain.
Answer: False

9. The supply of glucose to the brain must be continuous.
Answer: True

10. The crossing of fibers to the opposite site of the spinal cord is called decussation.
Answer: True

11. The satiety center is located in the medulla oblongata.
Answer: False

12. The central sulcus separates the temporal and parietal lobes.
    Answer: False

13. The auditory association area is located in the diencephalon.
    Answer: False

14. Dopamine is an excitatory neurotransmitter.
    Answer: True

15. The limbic system has both conscious and unconscious components.
    Answer: True

16. The dense connective tissue coverings on a neuron help prevent impulses from conducting from one neuron to an adjacent one.
    Answer: True

17. The epineurium covers each individual neuron.
    Answer: False

18. Muscle coordination is a function of the medulla oblongata.
    Answer: False

19. There are twelve pairs of cranial nerves and 31 pairs of spinal nerves.
    Answer: True.

20. Both cranial and cranial nerves have dorsal and anterior roots.
    Answer: False

Short Answer

Write the word or phrase that best completes each statement or answers the question.

1. An injection of an anesthetic into the epidural space results in a(n) ____.
   Answer: epidural block

2. An inflammation of the meninges is known as ____.
   Answer: meningitis

3. The axillary nerve originates from the _____ plexus.
   Answer: brachial

4. The patellar reflex is a(n) ____ reflex.
   Answer: somatic

5. Spinal nerves divide into several branches, known as ____.
   Answer: rami

6. The sciatic nerve arises from the ____ plexus.
   Answer: sacral

7. The spinothalamic tract is a(n) ____ tract.
   Answer: ascending

8. The fluid that circulates through the ventricles is the ____.
   Answer: cerebrospinal fluid (CSF)

9. The circulatory route at the base of the brain is called the ____.
   Answer: cerebral arterial circle or circle of Willis

10. The barrier between the blood and the brain is called ____.
    Answer: blood-brain barrier

11. Humans awaken and sleep in a fairly constant rhythm called a ____ rhythm.
    Answer: circadian

12. The primary olfactory area is located in the ____ lobe.
    Answer: temporal

13. Memory that lasts for only a few seconds is ____.
    Answer: short-term memory

14. Brain waves can be measured with the ____.
    Answer: electroencephalogram (EEG)

15. Nerves arising from the lowest portion of the spinal cord are called the ____.
    Answer: cauda equina

16. The center for thirst is located in the ____.
    Answer: hypothalamus

17. The roughly triangular structures in the medulla that carry motor impulses, are called ____.
    Answer: pyramids

18. The CSF has two functions related to homeostasis: protection and ____.
    Answer: circulation

19. The vagus nerve originates in the ____.
    Answer: medulla

20. A group of widely scattered neurons with a net-like appearance is referred to as ____.
    Answer: reticular formation

21. The gray matter of the cerebrum is found in the ____.
    Answer: cerebral cortex

22. The lenticular nucleus is subdivided into the globus pallidus and the ____.
    Answer: putamen

23. The pons is part of the ____ ____.
    Answer: brain stem

24. Cerebrospinal fluid is produced by the ____ ____.
    Answer: choroid plexus

25. Inactivation of the RAS produces _____.
    Answer: sleep

Matching

Choose the item from Column 2 that best matches each item in Column 1.

1. Column 1: Optic nerve
   Column 2: II

2. Column 1: Trochlear nerve
   Column 2: IV

3. Column 1: Abducens
   Column 2: VI

4. Column 1: Glossopharyngeal
   Column 2: IX

5. Column 1: Accessory
   Column 2: XI

6. Column 1: A cerebrovascular accident.
   Column 2: Stroke

7. Column 1: A disabling memory disorder.
   Column 2: Alzheimer's disease

8. Column 1: Paralysis of both lower limbs.
   Column 2: Paraplegia

9. Column 1: A group of motor disorders resulting in the loss of muscle contraction.
   Column 2: Cerebral palsy

10. Column 1: A disorder that involves dopamine.
    Column 2: Parkinson's disease

Essay

Write your answer in the space provided or on a separate sheet of paper.

1. Name and briefly describe the meninges.
   Answer: The dura mater is the outermost layer and consists of tough, dense, irregular connective tissue. The arachnoid is the middle layer arranged in delicate webs of collagen and elastic fibers. The inner layer is the pia mater, a transparent layer of collagen and elastic fibers.

2. Name the principal parts of the brain and their subdivisions.
   Answer: Cerebrum: cerebral cortex and cerebral white matter
   Diencephalon: thalamus and hypothalamus
   Brain stem: medulla, pons, and midbrain
   Cerebellum:

3. Name the brain ventricles and the structures which connect them.
   Answer: The two lateral ventricles are connected to the third ventricle by the interventricular foramen. The third ventricle is connected to the fourth ventricle by the cerebral aqueduct, and the fourth ventricle is continuous with the central canal of the spinal cord.

4. Describe the functions of the hypothalamus.
   Answer: The hypothalamus controls and integrates autonomic system activities, it controls the release of many hormones of the pituitary gland, it controls body temperature, and it regulates food intake. The hypothalamus is also associated with feelings such as aggression, pain, and pleasure. The thirst center is also found in the hypothalamus. Furthermore, the hypothalamus is one of the areas that maintains consciousness and sleep patterns.

CHAPTER 11 Autonomic Nervous System

Multiple-Choice

Choose the one alternative that best completes the statement or answers the question.

1. Which of the following statements about the autonomic nervous system is NOT true?
   A) it regulates the size of the pupils
   B) all of the fibers are afferent fibers
   C) it consists of two subdivisions
   D) it regulates visceral activities
   D) it is an involuntary system
   Answer: B

2. The first motor neuron in an autonomic pathway is called a
   A) preganglionic neuron.
   B) postganglionic neuron.
   C) autonomic ganglion.
   D) sympathetic neuron.
   E) parasympathetic neuron.
   Answer: A

3. The part of the nervous system that regulates smooth muscle is the
   A) somatic nervous system.
   B) limbic system.
   C) autonomic nervous system.
   D) Both A and B.
   E) Both B and C.
   Answer: C

4. The sympathetic division of the ANS is also referred to as the
   A) autonomic division.
   B) craniosacral division.
   C) thoracolumbar division.
   D) somatic division.
   E) ganglionic division.
   Answer: C

5. All of the following are autonomic ganglia EXCEPT
   A) sympathetic trunk ganglia.
   B) prevertebral ganglia.
   C) terminal ganglia.
   D) dorsal root ganglia.
   E) All of the above are autonomic ganglia.
   Answer: D

6. Which of the following statements does NOT apply to the autonomic nervous system?
   A) Autonomic pathways consist of sets of two motor neurons.
   B) ANS neurons always release acetylcholine as their neurotransmitter.
   C) The output part of the ANS has two principal branches.
   D) The postganglionic neuron lies entirely in the peripheral nervous system.
   E) All of the above are correct statements.
   Answer: B

7. Sympathetic fibers
   A) are unmyelinated fibers.
   B) originate in the brain.
   C) originate in the spinal cord.
   D) tend to be long at the preganglionic site.
   E) synapse in the terminal ganglion.
   Answer: C

8. All of the following are involved in the control of the autonomic nervous system EXCEPT
   A) the cerebral cortex.
   B) the cerebellum.
   C) the medulla oblongata.
   D) thalamus.
   E) the hypothalamus.
   Answer: B

9. All of the following are visceral effectors EXCEPT
   A) cardiac muscle.
   B) smooth muscle.
   C) skeletal muscle.
   D) glandular epithelium.
   E) all of the above are visceral effectors.
   Answer: C

10. The __ nervous system only contains all sensory neurons and all voluntary motor neurons.
   A) peripheral
   B) somatic
   C) autonomic.
   D) central.
   E) parasympathetic.
   Answer: B

11. Which of the following is true of BOTH the parasympathetic and sympathetic divisions of the autonomic nervous system?
   A) they have long preganglionic neurons.
   B) they have ganglia next to the effectors.
   C) they use acetylocholine as the first neurotransmitter.
   D) they use acetylocholine to excite the effector.
   E) they have long postganglionic neurons.
   Answer: C

12 In the autonomic nervous system there are ___ neurons between the CNS and the visceral effector.
   A) 5
   B) 4
   C) 3
   D) 2
   E) 1
   Answer: D

13. The preganglionic fibers of the autonomic nervous system release
   A) noradrenaline.
   B) norephinephrine.
   C) acethylcholine.
   D) GABA.
   E) adrenaline.
   Answer: C

14. Preganglionic fibers of the parasympathetic division of the ANS synapse with the
   A) sympathetic trunk ganglia.
   B) celiac ganglion.
   C) mesenteric ganglion.
   D) terminal ganglia.
   E) prevertebral ganglia.
   Answer: D

15. All of the following are true about preganglionic fibers of the sympathetic division EXCEPT
   A) they are myelinated.
   B) they leave the spinal cord through the anterior root.
   C) they synapse with many postganglionic cell bodies.
   D) they are longer than the postganglionic fibers.
   E) they use acetylcholine as the neurotransmitter.
   Answer: D

16. Adrenergic neurons are neurons that release
   A) acetylcholine.
   B) norepinephrine.
   C) histamine.
   D) Both A and B.
   E) All of the above.
   Answer: B

17. The sympathetic nervous system does all of the following EXCEPT
   A) speeds up digestion so there will be glucose for muscles.
   B) has sympathetic trunk ganglia.
   C) speeds up heart rate.
   D) speeds up respiratory rate.
   E) uses epinephrine as a neurotransmitter.
   Answer: A

18. The neurotransmitter of postganglionic fibers of the sympathetic division is
   A) norepinephrine.
   B) acetylcholine.
   C) acetylcholinesterase.
   D) dopamine.
   E) GABA.
   Answer: A

19. The neurotransmitter of postganglionic fibers in the parasympathetic division is
   A) glycine.
   B) acetylcholine.
   C) acetylcholinesterase.
   D) dopamine.
   E) norepinephrine.
   Answer: B

20. Impulses for one division of the ANS stimulates an organ's activity while the other division inhibits the organ's activity. This is called
    A) double innervation.
    B) dual innervation.
    C) autonomic innervation.
    D) synergistic innervation.
    E) agonistic innervation.
    Answer: B

21. The sympathetic division of the ANS
    A) increases the activities of the heart.
    B) decreases the activities of the heart.
    C) increases digestive processes.
    D) Both A and C.
    E) None of the above.
    Answer: A

22. All of the following are fight-or-flight responses EXCEPT
    A) increase of the heart rate.
    B) increase in blood pressure.
    C) constriction of the pupils.
    D) dilation of the bronchioles.
    E) decrease of digestive secretions.
    Answer: C

23. Somatic motor neurons
    A) have long preganglionic fibers.
    B) release norepinephrine.
    C) release acetylcholine.
    D) innervate smooth muscle.
    E) are regulated by the hypothalamus.
    Answer: C

24. The autonomic nervous system
    A) is under voluntary control.
    B) is always excitatory.
    C) has one motor neuron from the CNS to the effector.
    D) operates without conscious control.
    E) innervates skeletal muscles.
    Answer: D

25. The portion of the nervous system regulating smooth muscle activity is the
    A) central nervous system.
    B) sympathetic nervous system.
    C) autonomic nervous system.
    D) somatic nervous system.
    E) somatosensory system.
    Answer: C

26. The balance between sympathetic and parasympathetic activity is regulated by the
    A) hypothalamus.
    B) hippocampus.
    C) cerebellum.
    D) medulla.
    E) pituitary.
    Answer: A

27. Which of the following systems is responsible for skeletal muscle activity
    A) utonomic nervous system.
    B) sympathetic nervous system.
    C) parasympathetic nervous system.
    D) somatosensory system.
    E) somatic nervous system.
    Answer: E

28. Autonomic fibers are collectively called
    A) somatic fibers.
    B) visceral motor fibers.
    C) visceral afferent fibers.
    D) visceral sensory fibers.
    E) cholinergic fibers.
    Answer: B

29. The divisions of the autonomic nervous system are called
    A) sympathetic and cholinergic divisions.
    B) parasympathetic and adrenergic divisions.
    C) cholinergic and adrenergic divisions.
    D) sympathetic and parasympathetic divisions.
    E) somatic and parasympathetic divisions.
    Answer: D

30. The somatic and autonomic nervous system make up the ____ nervous system.
    A) involuntary
    B) peripheral
    C) central
    D) skeletal
    E) visceral
    Answer: B

31. Which of the following responses is a sympathetic response?
    A) Increased bile secretion
    B) Breakdown of triglycerides in adipose tissue.
    C) Vasodilation in the penis.
    D) Promotion of insulin secretion.
    E) All of the above are due to sympathetic stimulation.
    Answer: B

32. The ganglia which lie on either side of the backbone are the
    A) sympathetic trunk ganglia.
    B) terminal ganglia.
    C) dorsal root ganglia.
    D) prevertebral ganglia.
    E) superior mesenteric ganglia.
    Answer: A

33. All of the following are effects of the parasympathetic division EXCEPT
    A) conserving body energy.
    B) restoring of body energy.
    C) dilation of pupils.
    D) promoting digestion.
    E) increase in bile secretion.
    Answer: C

34. Insulin secretion is promoted by the activity of the
    A) sympathetic nervous system.
    B) hypothalamus.
    C) somatic nervous system.
    D) parasympathetic nervous system.
    E) autonomic ganglia.
    Answer: D

35. The adrenal glands produce epinephrine and norepinephrine to intensify and prolong ___ effects.
    A) CNS
    B) endocrine
    C) somatic
    D) parasympathetic
    E) sympathetic
    Answer: E

36. Acetylcholine is inactivated by
    A) pseudocholinesterase.
    B) acetylcholinesterase.
    C) acetase.
    D) norepinephrine.
    E) epinephrine.
    Answer: B

37. The acronym "SLUDD" deals with
    A) paradoxical fear.
    B) parasympathetic tone.
    C) "fight-or-flight" response.
    D) Both A and B.
    E) All of the above.
    Answer: D

38. Excessive contraction of arterioles within the fingers and toes due to prolonged sympathetic stimulation can be
    A) autonomic dysreflexia.
    B) Horner's syndrome.
    C) Down's syndrome.
    D) Raynaud's disease.
    E) Reye's syndrome.
    Answer: D

39. The parasympathetic nervous system decreases all of the following except
    A) heart rate.
    B) respiratory rate.
    C) digestion.
    D) pupil size.
    E) air pathway diameter.
    Answer: C

40. The effect of the parasympathetic nervous system on the eye is to
    A) help the eye move to focus directly on an object.
    B) decrease the secretion of tears.
    C) dilate the pupil.
    D) constrict the pupil.
    E) adjust the lens for far vision.
    Answer: D

True-False

Write T if the statement is true and F if the statement is false.

1. The hypothalamus receives input from areas of the nervous system concerned with emotions.
   Answer: True

2. The autonomic nervous system innervates the skeletal muscles.
   Answer: False

3. The parasympathetic division of the ANS originates in the brain and spinal cord.
   Answer: True

4. The sympathetic division is involved with energy expenditure.
   Answer: True

5. Fear stimulates the parasympathetic division of the ANS.
   Answer: False

6. The dilation of pupils is an autonomic response.
   Answer: True

7. All preganglionic fibers of the ANS are cholinergic.
   Answer: True

8. Terminal ganglia are sympathetic ganglia.
   Answer: False

9. Parasympathetic ganglia are near or within visceral effectors.
   Answer: True

10. Parasympathetic stimulation increases sweat secretion.
    Answer: False

11. The parasympathetic nervous system only originates from the cranial nerves.
    Answer: False.

12. The parasympathetic nervous system has long preganglionic axons and short postganglionic axons.
    Answer: True

13. Because of its connection to "fight-or-flight", increased sympathetic stimulation is connected to stress responses.
    Answer: True.

14. The parasympathetic branch of the autonomic nervous system prepares the body or the "fight-or-flight" response.
    Answer: False

15. Pre- and postganglionic fibers are present in the somatic nervous system.
    Answer: False

16. Both the parasympathetic and sympathetic divisions have two neurons leading to the effectors which use two different neurotransmitters.
    Answer: False

17. The effects of the adrenal medulla are linked to the activity of the parasympathetic nervous system.
    Answer: False

Short Answer

Write the word or phrase that best completes each statement or answers the question.

1. The subdivision of the autonomic nervous system are the ____ and ____ divisions.
   Answer: sympathetic and parasympathetic

2. Autonomic fibers are also called ____.
   Answer: visceral motor fibers

3. The sympathetic division is also called ____ division.
   Answer: thoracolumbar

4. The celiac ganglion is a ____ ganglion.
   Answer: sympathetic

5. The part of the nervous system that regulates smooth and cardiac muscle is the ______ nervous system.
   Answer: autonomic

6. Cholinergic fibers release ____.
   Answer: acetylcholine

7. The enzyme that inactivates acetylcholine is ____.
   Answer: acetylcholinesterase

8. Synapses between autonomic neurons and their effectors are called __________ junctions.
   Answer: neuroeffector

9. The dilation of bronchioles is due to ____ activity.
   Answer: sympathetic

10. The somatic nervous system is under ____ control.
    Answer: voluntary

11. The neurotransmitter of all preganglionic fibers is ____.
    Answer: acetylcholine

12. The neurotransmitter of sympathetic postganglionic fibers is ____.
    Answer: norepinephrine

13. The ____ division of the ANS stimulates epinephrine and norepinephrine secretion.
    Answer: sympathetic

14. The ____ division of the ANS inhibits digestion.
    Answer: sympathetic

15. All the axons of the ANS are ____ fibers.
    Answer: motor

16. Organs that receive impulses from both divisions of the ANS have ____.
Answer: dual innervation

17. Neurons that release acetylcholine are referred to as ___________ neurons.
Answer: cholinergic

18. Synapses of visceral motor fibers occur in ____.
Answer autonomic ganglia

19. Long preganglionic fibers belong to the ____ division.
Answer: parasympathetic

20. The neuroeffector junctions at glands are referred to as ____.
Answer: neuroglandular junctions

21. Parasympathetic impulses ____ the rate of the heartbeat.
Answer: decrease

22. During extreme stress the ____ division dominates the ____ division.
Answer: sympathetic, parasympathetic

23. The division that conserves and restores the body's energy is the _______ division.
Answer: parasympathetic

24. The acronym helpful in remembering five responses that occur when parasympathetic tone rises is "________".
Answer: SLUDD

25. Renin secretion in the kidney is a ___________ response.
Answer: sympathetic

Matching

Choose the item from Column 2 that best matches each item in Column 1.

1. Column 1: Decreases heart rate
Column 2: Parasympathetic

2. Column 1: Increases digestion
Column 2: Parasympathetic

3. Column 1: Stimulates renin secretion
Column 2: Sympathetic

4. Column 1: Dilation of pupils
Column 2: Sympathetic

5. Column 1: Stimulates salivary gland secretion
Column 2: Parasympathetic

6. Column 1: Inhibits gastric gland secretion
Column 2: Sympathetic

7. Column 1: Increased respiratory rate
   Column 2: Sympathetic

8. Column 1: Erection of hairs
   Column 2: Sympathetic

9. Column 1: Promotes gastric gland secretion
   Column 2: Parasympathetic

10. Column 1: Constricts pupils
    Column 2: Parasympathetic

Essay

Write your answer in the space provided or on a separate sheet of paper.

1. Name the divisions of the autonomic nervous system and compare their functions.
   Answer: The parasympathetic division of the ANS controls bodily functions during normal activities, while the sympathetic nervous system takes over during stress and prepares the body for the "fight-or-flight" response.

2. Compare the length of the pre- and postganglionic fibers of the ANS divisions and name the neurotransmitters used in each of the fibers.
   Answer: Parasympathetic division: long preganglionic fibers which use acetylcholine as neurotransmitter, and short postganglionic fibers which also use acetylcholine. Sympathetic division: short preganglionic fibers using acetylcholine, and long post ganglionic fibers utilizing noradrenaline.

3. Compare the functional differences between the autonomic and somatic nervous system.
   Answer: The somatic nervous system innervates skeletal muscles and is under voluntary control, the autonomic nervous system innervates cardiac muscle, smooth muscle, and glands. The ANS is an involuntary system.

CHAPTER 12 Somatic Senses and Special Senses

Multiple-Choice

Choose the one alternative that best completes the statement or answers the question.

1. The simplest sensory receptors are
   A) thermoreceptors.
   B) encapsulated nerve endings.
   C) proprioceptors.
   D) free nerve endings.
   E) touch receptors.
   Answer: D

2. A decreased sensitivity that occurs when a stimulus is continuously applied to sensory receptors is called
   A) adaptation.
   B) relaxation.
   C) accommodation.
   D) refraction.
   E) consolidation.
   Answer: A

3. Which of the following senses does NOT belong to the group of somatic senses?
   A) touch receptors
   B) temperature receptors
   C) visceral receptors
   D) pressure receptors
   E) proprioceptors
   Answer: C

4. All of the following are cutaneous sensations EXCEPT
   A) touch.
   B) pressure.
   C) temperature.
   D) equilibrium.
   E) vibration.
   Answer: D

5. All of the following are mechanoreceptors EXCEPT
   A) hair root plexuses.
   B) Meissner's corpuscles.
   C) nociceptors.
   D) Pacinian corpuscles.
   E) organs of Ruffini.
   Answer: C

6. Which of the following senses falls under the category of complex receptors?
   A) taste
   B) touch
   C) heat
   D) cold
   E) pain
   Answer: A

7. Touch receptors are classified as
   A) mechanoreceptors.
   B) nociceptors.
   C) chemoceptors.
   D) photoreceptors.
   E) proprioceptors.
   Answer: A

8. Pressure receptors, which are located around joints, tendons, and muscles, are
   A) Pacinian corpuscles.
   B) Meissner's corpuscle.
   C) organs of Ruffini.
   D) tactile receptors.
   E) nociceptors.
   Answer: A

9. The sensation of *itch* results from the stimulation of
   A) Meissner's corpuscles.
   B) Pacinian corpuscles.
   C) free nerve endings.
   D) proprioceptors.
   E) tactile receptors.
   Answer: C

10. A kind of pain often experienced by people suffering a heart attack is called
   A) referred pain.
   B) delayed pain.
   C) dull pain.
   D) sharp pain.
   E) phantom pain.
   Answer: A

11. The sense of muscle activity is called
   A) proprioception.
   B) nociception.
   C) stretch reception.
   D) adaptation.
   E) gustation.
   Answer: A

12. Muscle spindles are examples of
   A) free nerve endings.
   B) proprioceptors.
   C) Pacinian corpuscles.
   D) itch receptors.
   E) None of the above.
   Answer: B

13. Olfactory receptors
   A) are located in the nasal cavity.
   B) are stimulated by gaseous particles.
   C) adapt.
   D) have knob-shaped dendrites.
   E) All of the above are correct.
   Answer: E

14. Which of the following areas of the brain are involved in olfaction?
    A) temporal lobe
    B) hypothalamus
    C) limbic system
    D) none of the above
    E) all of the above
    Answer: E

15. Taste receptors are called
    A) taste buds.
    B) gustatory receptors.
    C) olfactory receptors.
    D) filiform papillae.
    E) taste pores.
    Answer: B

16. The region that is most sensitive to bitter taste is the
    A) tip of the tongue.
    B) back of the tongue.
    C) middle of the tongue.
    D) margin of the tongue.
    E) side of the tongue.
    Answer: B

17. The cranial nerves involved in the gustatory pathway are the
    A) glossopharyngeal, vagus, and gustatory.
    B) optic, glossopharyngeal, and facial.
    C) vagus, hypoglossal, and abducens.
    D) facial, glossopharyngeal, and vagus.
    E) vestibulochochlear, abducens, and hypoglossal.
    Answer: D

18. A physician specialized in the diagnosis and treatment of eye disorders with drugs, surgery, and corrective lenses is a(n)
    A) oncologist.
    B) pharmacologist.
    C) optician.
    D) ophthalmologist.
    E) optometrist.
    Answer: D

19. The immediate structure leading tears away from the lacrimal glands is/are the
    A) nasolacrimal duct.
    B) lacrimal ducts.
    C) eustachian tube.
    D) nasal cavity.
    E) nasal concha.
    Answer: B

20. The bactericidal enzymes present in tears are
    A) lysosomes.
    B) lysozyme.
    C) bacteriozyme.
    D) lipase.
    E) protease.
    Answer: B

21. The “white” of the eye is called
   A) conjunctiva.
   B) sclera.
   C) retina.
   D) cornea.
   E) iris.
   Answer: B

22. Which of the following belongs to the vascular tunic of the eye
   A) cornea.
   B) sclera.
   C) retina.
   D) iris.
   E) pupil.
   Answer: D

23. The muscle responsible for the change of the shape of the lens is the
   A) arrector pili muscle.
   B) orbicularis oculi.
   C) ciliary muscle.
   D) superior rectus muscle.
   E) retinal muscle.
   Answer: C

24. The layer that contains photoreceptors is the
   A) iris.
   B) retina.
   C) sclera.
   D) cornea.
   E) iris.
   Answer: B

25. The area containing the highest concentration of cones is the
   A) central fovea.
   B) optic disc.
   C) iris.
   D) macula lutea.
   E) pupil.
   Answer: A

26. The clear jelly-like substance behind the lens of the eye is the
   A) aqueous humor.
   B) cellular body.
   C) ciliary body.
   D) anterior humor.
   E) vitreous humor.
   Answer: E

27. The bending of light that happens at the cornea and lens is
   A) accommodation.
   B) reflection.
   C) contraction.
   D) refraction.
   E) adaptation.
   Answer: D

28. The inability to clearly see near objects is called
    A) nearsightedness.
    B) astigmatism.
    C) hypermetropia.
    D) presbyopia.
    E) myopia.
    Answer: C

29. The "canal of Schlemm" is another term for the
    A) vitreous chamber.
    B) vitreous body.
    C) anterior cavity.
    D) aqueous humor.
    E) scleral venous sinus.
    Answer: E

30. The photopigment in rods is
    A) chlorophyll.
    B) rhodopsin.
    C) opsin.
    D) melanin.
    E) retinal.
    Answer: B

31. Which of the following form the optic nerve
    A) ganglion cells.
    B) retinal cells.
    C) bipolar cells.
    D) rods.
    E) cones.
    Answer: A

32. All of the following belong to the external ear EXCEPT
    A) auditory tube.
    B) auricle.
    C) external auditory canal.
    D) tympanic membrane.
    E) ceruminous glands.
    Answer: A

33. Which of the following belongs to the middle ear?
    A) cochlea.
    B) ossicles.
    C) bony labyrinth.
    D) vestibule.
    E) scala tympani.
    Answer: B

34. The middle part of the bony labyrinth is the
    A) cochlea.
    B) saccule.
    C) utricle.
    D) vestibule.
    E) ampulla.
    Answer: D

35. Located between the cochlear duct and the scala vestibuli is the
A) vestibular membrane.
B) tectorial membrane.
C) basement membrane.
D) cochlear duct.
E) basilar membrane.
Answer: A

36. The center of the eardrum is connected to the
A) incus.
B) cochlea.
C) vestibule.
D) malleus.
E) staples.
Answer: D

37. The receptors for hearing are located in the
A) semicircular canals.
B) cochlea.
C) utricle.
D) sacculae.
E) vestibule.
Answer: B

38. The senses for dynamic equilibrium are located in the
A) utricle.
B) sacculae.
C) Organ of Corti.
D) cochlea.
E) semicircular ducts.
Answer: E

39. An abnormally high intraocular pressure is referred to as
A) cataract.
B) glaucoma.
C) trachoma.
D) vertigo.
E) nystagmus.
Answer: B

40. The most common cause of blindness in the United States is
A) scotoma.
B) trachoma.
C) cataracts.
D) glaucoma.
E) macular degeneration.
Answer: D

41. "Pinkeye" is the common term for
A) trachoma.
B) conjunctivitis.
C) myringitis.
D) blepharitis.
E) keratitis.
Answer: B

42. If you enter a room and smell a strong odor which seems to soon fade away, you have experienced
   A) damage to your sensory receptors.
   B) sensory adaption.
   C) pressure on chemoreceptors.
   D) referred pain.
   E) a neuronal loss in the olfactory center of the frontal lobe.
   Answer: B

43. When you sniff to help you detect or identify odors, you are
   A) increasing the pressure on the mechanoreceptors.
   B) forcing air to the back of the throat toward the olfactory nerves.
   C) forcing air to the top of the nasal cavity where the chemoreceptors are located.
   D) forcing gasses into solution.
   Answer: C

44. The sense of balance that tells you "where your head is at" when you are not moving is called ____ equilibrium and is a function of the __ in the inner ear.
   A) static & cochlear duct.
   B) dynamic & semicircular canals.
   C) dynamic & utricle and saccule.
   D) static & utricle and saccule.
   E) static & semicircular canals.
   Answer: D

45. What is the function of the extrinsic eye muscles?
   A) change the size of the pupil for light and dark conditions.
   B) change the shape of the lens for near or far conditions.
   C) turn the auricle towards the source of the sound.
   D) increase or decrease the amount of aqueous humor.
   E) roll the eyes after taking a particularly awful test.
   Answer: E

46. Which of the following is abnormal in someone who is unable to see in low light (night blind)?
   A) rods.
   B) cones.
   C) cornea.
   D) choroids coat.
   E) optic nerve.
   Answer: A

47. The ossicles of the ear are responsible for
   A) sound conduction and amplification.
   B) static equilibrium.
   C) dynamic equilibrium.
   D) pressure equalization.
   E) conversion of waves into electrical impulses.
   Answer: A

48 Accommodation involves changing the shape of the
   A)cornea.
   B) stapes.
   C) lens.
   D) utricle and saccule.
   E) retina.
   Answer: B

49. Which of these changes vibrations to nerve impulses?
    A) cochlea and Organ of Corti.
    B) stapes and oval window.
    C) utricle and saccule.
    D) tympanic membrane,
    E) retine.
    Answer: A

50. Referred pain occurs because of
    A) shared pathways between visceral organs and body surfaces.
    B) adaptation of nerve pathways.
    C) lack of myelination on pain neurons.
    D) the autonomic nervous system.
    E) all of the above.
    Answer: A

True-False

Write T if the statement is true and F if the statement is false.

1. Muscle spindles are found at the junction of a tendon with a muscle.
   Answer: False

2. Equilibrium is classified as a special sense.
   Answer: True

3. Cutaneous receptors consist of the axons of sensory neurons.
   Answer: False

4. Crude touch refers to the feeling that something has touched the skin and the exact location can be determined.
   Answer: False

5. Thermoreceptors are free nerve endings.
   Answer: True

6. Pain is essential because it protects us from greater damage.
   Answer: True

7. The highest concentration of taste buds is found in filiform papillae.
   Answer: False

8. The adult eyeball measures about 3.5 cm in diameter.
   Answer: False

9. The middle layer of the eyeball is the vascular tunic.
   Answer: True

10. Cones are specialized for color vision.
    Answer: True

11. Night blindness is most often caused by a vitamin C deficiency.
    Answer: False

12. Some fibers of the optic nerve cross in the optic tract.
    Answer: False

13. The middle ear is a small air-filled cavity between the eardrum and the internal ear.
    Answer: True

14. The walls of the utricle contain a small, flat region called macula.
    Answer: True

15. Almost any excessive sensation such as loud noise, can cause the sensation of pain.
    Answer: True.

16. Sensation of senses actually occurs in the brain. In other words, perception occurs when the brain interprets what the electrical impulse from the receptor is trying to tell it.
    Answer: True

17. Sound is conducted through air, bone and fluid before it is changed to electrical impulses in the inner ear.
    Answer: True

18. The middle tunic of the eye consists of the choroids coat, cornea, lens and ciliary body.
    Answer: False

19. When the ciliary muscle contracts, the lens flattens for far vision.
    Answer: False

20. The muscles of the iris and ciliary muscle are considered to be extrinsic eye muscles because they are voluntary.
    Answer: False

Short Answer

Write the word or phrase that best completes each statement or answers the question.

1. The depolarization of the membrane in sensory receptors is called ______ _________.
   Answer: generator potential

2. The small area of the retina, which does not contain photoreceptors but the optic nerve, is the ____.
   Answer: optic disc

3. Receptors that detect pressure or stretching are collectively called ____________.
   Answer: mechanoreceptors

4 The conscious or subconscious awareness of external or internal conditions of the body is a(n) _____.
   Answer: sensation

5. Conscious sensations are integrated in the _________ _________.
   Answer: cerebral cortex

6. The ability to recognize which point of the body is touched is referred to as ____.
   Answer: discriminative touch

7. Receptors for pain are called ____.
   Answer: nociceptors

8. Receptors for taste are called ____ receptors.
Answers: gustatory

9. The ___ gland produces tears.
Answer: lacrimal

10. The colored portion of the eyeball is the ____.
Answer: iris

11. A loss of transparency of the lens is known as ____.
Answer: cataract

12. The ability of the lens to change its curvature is called ____.
Answer: accommodation

13. Images that focus upside down on the retina are called ____ images.
Answer: inverted

14. The receptors for color vision and bright light are ____.
Answer: cones

15. ____ glands produce earwax.
Answer: ceruminous

16. The bony labyrinth contains fluid called the ____.
Answer: perilymph

17. The organ of hearing is the ____.
Answer: Organ of Corti (spiral organ)

18. The balance and posture of the body without movement of the head is called ____.
Answer: static equilibrium

19. An inflammation of the conjunctiva is called ____.
Answer: conjunctivitis

20. The opening of a taste bud is the ____.
Answer: taste pore

21. The hole in the center of the iris is the ____.
Answer: pupil

22. The pressure in the eye that is mainly produced by the aqueous humor is the ____.
Answer: intraocular pressure

23. The thin membrane between the external auditory canal and the ossicles is the ____.
Answer: tympanic membrane

24. The greater the frequency of sound the higher the ____.
Answer: pitch

25. A type of receptor that provides information about the internal environment is called a(n) _______.
Answer: interoceptor

Matching

Choose the item from Column 2 that best matches each item in Column 1.

1. Column 1: An acute bacterial infection of the middle ear.
   Column 2: otitis media

2. Column 1: A disorder brought on by motion.
   Column 2: motion sickness

3. Column 1: An inflammation of the eyelid.
   Column 2: blepharitis

4. Column 1: An inflammation of the auditory tube.
   Column 2: eustachitis

5. Column 1: An inflammation of the inner ear.
   Column 2: labyrinthitis

6. Column 1: A rapid involuntary movement of the eyeballs.
   Column 2: nystagmus

7. Column 1: Dilated pupil
   Column 2: mydriasis

8. Column 1: A ringing in the ears.
   Column 2: tinnitus

9. Column 1: A change in the environment capable of activating sensory neurons.
   Column 2: stimulus

10. Column 1: The receptors of pain.
    Column 2: nociceptors

11. Column 1: A pain in the skin overlying the stimulated organ.
    Column 2: referred pain

12. Column 1: The lining of the eyelid.
    Column 2: conjunctiva

13. Column 1: Responsible for night vision.
    Column 2: Rods

14. Column 1: Responsible for dynamic equilibrium.
    Column 2: semicircular ducts

15. Column 1: Corpuscles of touch.
    Column 2: Meissner's corpuscles

Essay

Write your answer in the space provided or on a separate sheet of paper.

1. Define the term sensation.
   Answer: Sensation refers to the awareness of external or internal conditions of the body.

2. Name the proprioceptive receptors and briefly describe the location of each.
   Answer: Muscle spindles: located between skeletal muscle fibers. Tendon organs: located at the junction of a tendon with muscle. Join kinesthetic receptors: located around synovial joints. Maculae and Cristae: located in the inner ear.

3. Describe the olfactory pathway.
   Answer: Axons of the olfactory receptors form the olfactory nerves, the olfactory nerves convey impulses to the olfactory bulbs, and then to the olfactory tract, which sends the impulses to the primary olfactory area in the temporal lobe of the cerebral cortex.

4. Name the three coats of the eye and structures that belong to each coat.
   Answer: Fibrous tunic: sclera and cornea. Vascular tunic: choroids, ciliary body, and iris. Retina: photoreceptor cells (rods and cones), bipolar cells, ganglion cells.

CHAPTER 13 The Endocrine System

Multiple-Choice

Choose the one alternative that best completes the statement or answers the question.

1. Which of the following systems work closely with the endocrine system to coordinate the body's functions?
   A) Cardiovascular system
   B) Respiratory system
   C) Reproductive system
   D) Nervous system
   E) Lymphatic system
   Answer: D

2. All of the following are endocrine glands EXCEPT:
   A) adrenal glands.
   B) sebaceous glands.
   C) parathyroid glands.
   D) pineal glands.
   E) pituitary glands.
   Answer: B

3. Which of the following organs contain hormone producing cells?
   A) stomach
   B) liver
   C) skin
   D) All of the above.
   E) None of the above.
   Answer: D

4. A second messenger in the action of many water-soluble hormones such as peptides is
   A) tRNA.
   B) ATP.
   C) cAMP.
   D) PTH.
   E) CoA.
   Answer: C

5. Hormones that bind to receptors within the target cells are
   A) lipid soluble.
   B) water soluble.
   C) second messengers.
   D) proteins.
   E) prohormones.
   Answer: A

6. The hormones regulating blood calcium levels are
   A) insulin and glucagon.
   B) glycogen and PTH.
   C) inhibiting hormones.
   D) PTH and calcitonin.
   E) calcitonin and ATP.
   Answer: D

7. A hormone that under certain circumstances is regulated by positive feedback is
   A) calcitonin.
   B) histamine.
   C) oxytocin.
   D) melatonin.
   E) insulin.
   Answer: C

8. The pituitary gland is attached to the hypothalamus by the
   A) epithalamus.
   B) infundibulum.
   C) parafollicular cells.
   D) intermediate mass.
   E) corpus callosum.
   Answer: B

9. Which of the following hormones stimulates testosterone production by the testis?
   A) TSH
   B) FSH
   C) ACTH
   D) LH
   E) GH
   Answer: D

10. Which of the following hormones is released in response to a nerve impulse?
    A) epinephrine
    B) cortisol
    C) testosterone
    D) insulin
    E) glucagon
    Answer: A

11. All of the following are hormones of the anterior pituitary EXCEPT
    A) human growth hormone.
    B) follicle-stimulating hormone.
    C) adrenocorticotropic hormone.
    D) prolactin.
    E) oxytocin.
    Answer: E

12. The hormone which effects skin pigmentation is
    A) MSH.
    B) FSH.
    C) GH.
    D) LH.
    E) ACTH.
    Answer: A

13. Which of the following hormones is released in response to hypoglycemia?
    A) TSH
    B) FSH
    C) LH
    D) GH
    E) MSH
    Answer: D

14. Insulinlike growth factor is released in response to
   A) insulin.
   B) glucagon.
   C) human growth hormone.
   D) luteneizing hormone.
   E) all pituitary hormones.
   Answer: C

15. All of the following hormones are released in response to releasing hormones EXCEPT
   A) thyroid stimulating hormone.
   B) human growth hormone.
   C) follicle stimulating hormone.
   D) prolactin.
   E) oxytocin.
   Answer: E

16. Which of the following hormones do neurosecretory cells produce?
   A) antidiuretic hormone
   B) calcitonin
   C) insulin
   D) growth hormone
   E) adrenocorticotropic hormone
   Answer: A

17. Antidiuretic hormone and oxytocin are produced by the
   A) posterior pituitary.
   B) anterior pituitary.
   C) thyroid.
   D) hypothalamus.
   E) pancreas.
Answer: D

18. The hormone which is released in large quantities just prior to childbirth is
   A) estrogen.
   B) oxytocin.
   C) prolactin.
   D) progesterone.
   E) growth hormone.
   Answer: B

19. Which of the following hormones controls the production and release of glucocoticoids?
   A) ADH
   B) ACTH
   C) GH
   D) FSH
   E) LH
   Answer: B

20. Calcitonin is a hormone of the
   A) adrenal cortex.
   B) thyroid gland.
   C) parathyroid gland.
   D) pituitary gland.
   E) thymus gland.
   Answer: B

21. Which of the following cells produce thyroxine?
   A) alpha cells
   B) oxyphil cells
   C) neurosecretory cells
   D) follicular cells
   E) parafollicular cells
   Answer: D

22. The hormone released in response to low blood calcium levels is
   A) oxytocin.
   B) parathyroid hormone.
   C) thyroxine.
   D) growth hormone.
   E) calcitonin.
   Answer: B

23. The hormone that inhibits the action of osteoclasts is
   A) oxytocin.
   B) vasopressin.
   C) calcitonin.
   D) parathyroid hormone.
   E) thyroxine.
   Answer: C

24. Mineralcorticoids
   A) are produced in the adrenal cortex.
   B) are steroid hormones.
   C) help regulate the homeostasis of sodium and potassium.
   D) All of the above.
   E) None of the above.
   Answer: D

25. Glucocorticoids are steroid hormones produced by the
   A) ovaries.
   B) testis.
   C) adrenal gland.
   D) thyroid gland.
   E) hypothalamus.
   Answer: C

26. Which of the following endocrine glands is directly controlled by the autonomic nervous system?
   A) adrenal medulla
   B) adrenal cortex
   C) anterior pituitary
   D) thyroid gland
   E) thymus gland
   Answer: A

27. Which of the following hormones are responsible for the fight-or-flight response?
   A) epinephrine
   B) norepinephrine
   C) acetylcholine
   D) A and B
   E) B and C
   Answer: D

28. The gland which can be classified as an endocrine and an exocrine glands is the
A) thyroid.
B) thymus.
C) pancreas.
D) pituitary.
E) hypothalamus.
Answer: C

29. Insulin is secreted by
A) alpha cells.
B) beta cells.
C) delta cells.
D) F-cells.
E) chief cells.
Answer: B

30. Glucagon
A) accelerates the conversion of glycogen into glucose.
B) slows down glucose formation from lactic acid.
C) decreases the conversion of glycogen into glucose.
D) speeds up protein synthesis within cells.
E) accelerates the transport of glucose from blood into cells.
Answer: A

31. The development and maintenance of the female sex characteristics is the responsibility of
A) estrogen and androgen.
B) progesterone and testosterone.
C) relaxin and inhibin.
D) progesterone and relaxin.
E) progesterone and estrogen.
Answer: E

32. FSH secretion is inhibited by
A) relaxin.
B) testosterone.
C) LH.
D) inhibin.
E) androgen.
Answer: D

33. Melatonin is a hormone of the
A) pituitary gland.
B) pancreas.
C) pineal gland.
D) thymus gland.
E) adrenal gland.
Answer: C

34. Prostaglandins
A) act as local hormones.
B) help induce inflammation.
C) are important in fat metabolism.
D) All of the above.
E) None of the above.
Answer: D

35. The second stage of the stress response is
   A) the stressor. Increases.
   B) the resistance reaction.
   C) the alarm reaction.
   D) the flight-or-fight response.
   E) the stage of exhaustion.
   Answer: B

36. Oversecretion of hGH during childhood is called
   A) giantism.
   B) acromegaly.
   C) pituitary dwarfism.
   D) cretinism.
   E) goiter.
   Answer: A

37. Insufficient ADH release causes
   A) diabetes insipidus.
   B) diabetes mellitus.
   C) tetany.
   D) cretinism.
   E) aldosteronism.
   Answer: A

38. Cushing's syndrome is due to oversecretion of
   A) glucocorticoids.
   B) testosterone.
   C) ADH.
   D) mineralcorticoids.
   E) thyroxine.
   Answer: A

39. A decline in the number of pancreatic beta cells is characteristic for
   A) diabetes insipidus.
   B) type II diabetes.
   C) type I diabetes.
   D) maturity onset diabetes.
   E) noninsulin-dependent diabetes.
   Answer: C

40. Cells that respond to a particular hormone are called
   A) receptor cells.
   B) effector cells.
   C) secretory cells.
   D) target cells.
   E) active cells.
   Answer: D

41. Hormone secretion is controlled by which of the following?
   A) nerve impulses.
   B) circulating chemicals.
   C) releasing hormones.
   D) inhibiting hormones.
   E) All of the above.
   Answer: E

42. The activities of the adrenal cortex are controlled by
   A) FSH.
   B) LH.
   C) hGH.
   D) MSH.
   E) ACTH.
   Answer: E

43. Which hormone stimulates cells to grow and divide?
   A) Thyroid stimulating hormone
   B) Luteinizing hormone
   C) Growth Hormone
   D) Gluccocorticoids
   E) Insulin
   Answer: C

44. Glucagon is produced in the ____ cells of the _____ .
   A) Beta, Islets of Langerhans
   B) Alpha, pancreas
   C) Beta, adrenal cortex
   D) Gamma, adrenal medulla
   E) None of the above.
   Answer: B

45. Insulin is secreted by the pancreas after a meal in order to:
   A) decrease the concentration of blood glucose.
   B) decrease the permeability of the cell membranes to glucose.
   C) increase the production of glucose from glycogen.
   D) increase the concentration of blood glucose.
   E) increase the amount of thyroid hormones in the blood.
   Answer: A

46. An adult who over secretes GH doesn't get taller. Why not?
   A) GH doesn't affect bones.
   B) Epiphyseal plates are closed in adults.
   C) Adult bones have dead cells
   D) The adult dies too soon to grow.
   E) Adults can't secrete GH.
   Answer: B

47. Under normal circumstances, individuals do not produce too much thyroxine because
   A) they run out of iodine.
   B) it is under negative feedback control.
   C) It gets turned into triiodothyronine.
   D) The thyroid is incapable of producing too much.
   E) All of the above.
   Answer: B

True-False

Write T if the statement is true and F if the statement is false.

1. The nervous system and the endocrine system regulate homeostasis.
   Answer: True

2. Hormones do not regulate the activity of the immune system.
Answer: False

3. Steroid hormones are derivatives of cholesterol.
Answer: True

4. Most peptide and protein hormones diffuse through the phospholipid bilayer to get into a target cell.
Answer: False

5. Hormone secretion is primarily regulated by negative feedback mechanisms.
Answer: True

6. Cells of the posterior pituitary produce vasopressin.
Answer: False

7. Some glands produce hormones in response to a signal from the autonomic nervous system.
Answer: True

8. Prolactin is a hormone involved in the initiation of milk production in the mammary glands.
Answer: True

9. Hormones of the hypothalamus control the release of human growth hormone from the anterior pituitary.
Answer: True

10. FSH stimulates the testis to produce testosterone.
Answer: False

11. Vasopressin is another name for antidiuretic hormone.
Answer: True

12. Alcohol inhibits ADH secretion resulting in an increased urine output.
Answer: True

13. Calcitonin is a hormone of the parathyroid gland.
Answer: False

14. The renin-angiotensin pathway is involved in the control of aldosterone secretion.
Answer: True

15. Androgens are hormones secreted by the male endocrine system only.
Answer: False

16. Steroid hormones act by binding to a protein receptor on the cell membrane of their target cell.
Answer: False.

17. The hypothalamus controls secretion of hormones from the anterior pituitary gland by nervous impulses.
Answer: False.

18. Some glands produce hormones in response to a signal from the autonomic nervous system.
Answer: True.

19. A player in the NBA would benefit from an injection of GH.
Answer: False.

20. All of the hormones of the pituitary gland and actually produced there.
Answer: False.

Short Answer

Write the word or phrase that best completes each statement or answers the question.

1. The science concerned with the endocrine system is called ____.
Answer: endocrinology

2. The endocrine system releases messenger molecules called ____.
Answer: hormones

3. Hormones that pass from secretory cells into the interstitial fluid and then into the blood are called ________ hormones.
Answer: circulating

4. The hypothalamus controls the activity of the anterior pituitary by way of __________.
Answer: releasing and inhibiting hormones

5. The production of oocytes by the ovaries is stimulated by ____.
Answer: follicle stimulating hormone (FSH)

6. Calcitriol is a hormone of the __________.
Answer: kidneys

7. Hormones that can enter the plasma membrane without a receptor are _____-_____ hormones.
Answer: lipid-soluble

8. The production and secretion of thyroid hormone is stimulated by ____.
Answer: thyroid stimulating hormone (TSH)

9. The formation of corpus luteum is stimulated by ____.
Answer: luteinizing hormone (LH)

10. The release of milk by the mammary glands depends on the hormone ____.
Answer: oxytocin

11. The receptors in the hypothalamus that detect low water concentration in the blood are ____.
Answer: osmoreceptors

12. The inhibition of bone breakdown is the function of ____.
Answer: calcitonin

13. Oxyphyl cells are found in the ____ gland.
Answer: parathyroid

14. The class of adrenal cortex hormones that deals with metabolism and resistance to stress are ____.
Answer: glucocorticoids

15. The two principal hormones of the adrenal medulla are ____ and ____.
Answer: epinephrine and norepinephrine

16. Mineralcorticoids are produced in the _______ _____.
    Answer: adrenal cortex

17. The endocrine portion of the pancreas consists of clusters of cells called ____.
    Answer: Islets of Langerhans (pancreatic islets)

18. The conversion of glucose into glycogen is accelerated by ____.
    Answer: insulin

19. Oversecretion of hGH during childhood results in ____.
    Answer: giantism

20. An enlarged thyroid gland is called ____.
    Answer: goiter

21. Hormones that influence other endocrine glands are called ____ hormones.
    Answer: tropic

22. Oxytocin and antidiuretic hormone are produced in cells of the ____.
    Answer: hypothalamus

23 Thyroid hormones regulate growth and development, the activity of the nervous system, and ____.
    Answer: metabolism

24. The adrenal gland is composed of the adrenal medulla and the ____.
    Answer: adrenal cortex

Matching

Choose the item from Column 2 that best matches each item in Column 1.

1. Column 1: Help induce inflammation.
   Column 2: prostaglandins

2. Column 1: Controls general body growth and metabolism.
   Column 2: hGH

3. Column 1: Stimulates the adrenal cortex.
   Column 2: ACTH

4. Column 1: Stimulates the development of the ovarian follicles.
   Column 2: FSH

5. Column 1: Initiates and maintains milk production by the mammary glands.
   Column 2: prolactin

6. Column 1: Decreases urine volume.
   Column 2: ADH

7. Column 1: Contains iodine.
   Column 2: thyroid hormone

8. Column 1: Are anti-inflammatory compounds.
   Column 2: glucocorticoids

9. Column 1: Secreted by alpha cells.
   Column 2: glucagon

10. Column 1: Secreted by the pineal gland.
    Column 2: melatonin

Essay

Write your answer in the space provided or on a separate sheet of paper.

1. Name the different chemical classes of hormones and give three examples of each.
   Answer: 1. Lipid derivatives: steroid hormones (aldosterone, cortisol, androgens, testosterone, estrogen, and progesterone); eicosanoids (prostaglandins, leukotrienes).
   2. Amino acid derivatives: thyroid hormones epinephrine, norepinephrine, histamine, serotonin, and melatonin.
   3. Peptides and proteins: releasing and inhibiting hormones, oxytocin, antidiuretic hormone, anterior pituitary hormones, insulin, glucagons, parathyroid hormone, calcitonin….

2. Name the seven broad areas of hormone effects.
   Answer: 1. Regulation of the chemical composition and volume of internal environment.
   2. Regulation of metabolism and energy balance.
   3. Regulation of smooth muscle and cardiac muscle contraction.
   4. Maintaining of homeostasis.
   5. Regulation of certain immune system activities.
   6. Regulation of growth and development.
   7. Regulation of reproductive activities.

3. Name the two basic mechanisms by which hormones can act on their target cells.
   Answer: Lipid-soluble hormones enter the target cell and bind to a receptor within the cell. The activated receptor will then alter cell function via the genes. Water-soluble hormones will bind to a membrane receptor of the target cell and activate a second messenger that is cyclic AMP.
   This activation will eventually produce the physiological response of the target cell.

CHAPTER 14 The Cardiovascular System: Blood

Multiple-Choice

Choose the one alternative that best completes the statement or answers the question.

1. Which of the following are functions of the blood?
   A) transportation
   B) regulation
   C) protection
   D) A and B
   E) All of the above
   Answer: E

2. The blood volume of an averaged sized male is
   A) 3 to 4 liters.
   B) 4 to 5 liters.
   C) 5 to 6 liters.
   D) 6 to 7 liters.
   E) 6 liters exactly.
   Answer: C

3. The pH range for blood is
   A) 7.45 – 7.65.
   B) 7.35 – 7.45.
   C) 7.25 – 7.35.
   D) 7.15 – 7.45.
   E) 7.00 – 7.35.
   Answer: B

4. The thin layer of white blood cells in centrifuged blood is called
   A) hematocrit.
   B) plasma.
   C) serum.
   D) buffy coat.
   E) matrix of blood.
   Answer: D

5. Most of the plasma proteins are
   A) albumins.
   B) fibrinogens.
   C) gamma globulins.
   D) alpha globulins.
   E) beta globulins.
   Answer: A

6. Which of the following belongs to agranular leukocytes?
   A) thrombocyte
   B) neutrophil
   C) basophil
   D) platelet
   E) monocyte
   Answer: E

7. Which of the following cells do NOT have a nucleus?
   A) erythrocytes
   B) granulocytes
   C) leukocytes
   D) monocytes
   E) agranulocytes
   Answer: A

8. The life span of red blood cells is
   A) 150 days.
   B) 140 days.
   C) 130 days.
   D) 120 days.
   E) 100 days.
   Answer: D

9. The pigment in red blood cells that carries oxygen is
   A) erythropoietin.
   B) melatonin.
   C) hemoglobin.
   D) uribilonogen.
   E) biliverdin.
   Answer: C

10. Worn-out red blood cells are phagocytized in the
   A) liver.
   B) spleen.
   C) red bone marrow.
   D) All of the above.
   E) None of the above.
   Answer: D

11. Which of the following components of hemoglobin can be reused by other cells for protein synthesis?
   A) the heme group
   B) globin
   C) transferin
   D) iron
   E) stercobilin
   Answer: B

12. Cellular oxygen deficiency is called
   A) ischemia.
   B) hypoxia.
   C) jaundice.
   D) anemia.
   E) cyanosis.
   Answer: B

13. What is the substance some athletes use for blood "doping?"
   A) hemoglobin
   B) iron
   C) erythropoietin
   D) transferrin
   E) adrenaline
   Answer: C

14. The first phagocytotic cell at the site of a bacterial invasion are
   A) lymphocytes.
   B) neutrophils.
   C) eosinophils.
   D) thrombocytes.
   E) monocytes.
   Answer: B

15. An allergic condition or a parasite infection is often manifested in a high count of
   A) eosinophils.
   B) neutrophils.
   C) basophils.
   D) monocytes.
   E) lymphocytes.
   Answer: A

16. Which of the following blood cells can develop into "wandering macrophages?
   A) neutrophils
   B) eosinophils
   C) basophils
   D) lymphocytes
   E) monocytes
   Answer: E

17. An increase in the number of white blood cells is called
   A) anemia.
   B) leukopenia.
   C) leukemia.
   D) leukocytosis.
   E) mononucleosis.
   Answer: D

18. Abnormally low levels of white blood cells which may be caused by radiation is called
   A) leukopenia.
   B) anemia.
   C) leukocytosis.
   D) leukemia.
   E) polycythemia.
   Answer: A

19. How many platelets should appear in each μL of blood?
   A) 20,000 – 40,000
   B) 100,000 – 150,000
   C) 200,000 – 250,000
   D) 150,000 – 400,000
   E) 400,000 – 600,000
   Answer: D

20. Which of the following blood proteins functions in blood-clotting?
   A) albumin
   B) fibrinogen
   C) alpha globulin
   D) beta globulin
   E) gamma globulin
   Answer: B

21. Which of the following sequences correctly describes the steps in the formation of a platelet plug?
   A) platelet release reaction, platelet aggregation, platelet adhesion
   B) platelet aggregation, platelet adhesion, platelet release reaction
   C) platelet adhesion, platelet release reaction, platelet aggregation
   D) platelet release reaction, platelet adhesion, platelet aggregation
   E) platelet adhesion, platelet aggregation, platelet release reaction
   Answer: C

22. The process of clotting in an unbroken blood vessel is called
   A) hemostasis.
   B) coagulation.
   C) thrombosis.
   D) vascular spasm.
   E) fibinolysis.
   Answer: C

23. Stoppage of bleeding is called
   A) hemostasis.
   B) vascular spasm.
   C) thrombosis.
   D) coagulation.
   E) embolism.
   Answer: A

24. The threads of a blood clot are formed by
   A) thrombin.
   B) prothrombin activator.
   C) platelet plug.
   D) fibrinogen.
   E) fibrin.
   Answer: E

25. All of the following are involved in blood clotting EXCEPT
   A) globulin.
   B) thrombin.
   C) calcium.
   D) prothrombinase.
   E) fibrin.
   Answer: A

26. Heredity deficiencies of coagulation is referred to as
   A) anemia.
   B) hemophilia.
   C) hemolysis.
   D) polycythemia.
   E) leukemia.
   Answer: B

27. The enzyme responsible for breaking up a blood clot is
   A) fibrinogen.
   B) ATPase.
   C) coagulase.
   D) plasminogen.
   E) plasmin.
   Answer: E

28. The anticoagulant produced by mast cells is
   A) plasminogen.
   B) plasmin.
   C) fibrinogen.
   D) heparin.
   E) vitamin K.
   Answer: D

29. A blood clot transported by the blood stream is a(n)
   A) platelet plug.
   B) thrombus.
   C) embolus.
   D) thrombin clot.
   E) None of the above.
   Answer: C

30. A person with blood type A has
   A) B antigens on the red blood cells.
   B) A antibodies in the plasma.
   C) A antigens on the red blood cells.
   D) Rh antigen on the red blood cells.
   E) A and B antibodies in the plasma.
   Answer: C

31. An individual with which ABO blood type can theoretically donate blood to recipients of all ABO types
   A) type A.
   B) type B.
   C) AB.
   D) type O.
   E) All of the above.
   Answer: D

32. The clumping of red blood cells due to incompatible blood transfusion is called
   A) hemolysis.
   B) embolism.
   C) agglutination.
   D) reaction.
   E) infiltration.
   Answer: C

33. Small inappropriate blood clots inside of a blood vessel are dissolved in a process called
   A) fibrinolysis.
   B) plasmolysis.
   C) clot retraction.
   D) thrombosis.
   E) agglutination.
   Answer: A

34. Insufficient production of erythrocytes due to lack of vitamin B12 is the cause for
   A) polycythemia.
   B) leukemia.
   C) aplastic anemia.
   D) pericious anemia.
   E) hemorrhagic anemia.
   Answer: D

35. The iron-containing portion of the hemoglobin molecule is
    A) ferrin.
    B) globin.
    C) heme.
    D) globulin.
    E) oxygen.
    Answer: C

36. The hormone released by the kidney and is stimulating RBC production is
    A) erythropoietin.
    B) plasmin.
    C) albumin.
    D) hemoglobin.
    E) histamine.
    Answer: A

37. The ion, which is essential in the blood-clotting mechanism, is
    A) magnesium.
    B) potassium.
    C) calcium.
    D) sodium.
    E) chlorine.
    Answer: C

38. The percentage of red blood cells in whole blood is called the
    A) hemoglobin content.
    B) platelet count.
    C) buffy coat.
    D) hematocrit.
    E) serum content.
    Answer: D

39. Anemia may be caused by
    A) bleeding.
    B) lack of sufficient nutrients in the diet.
    C) breakdown of RBC's prematurely.
    D) A & B above
    E) All of the above.
    Answer: E

40. Which of the following contain hemoglobin?
    A) eosinophils
    B) erythrocytes
    C) leukocytes
    D) thrombocytes
    E) platelets
    Answer: B

41. Dracula has the B agglutinin but not the A antigen, what blood type does he have?
    A) A
    B) B
    C) AB
    D) O
    E) Rh+
    Answer: D

42. Any decrease in the oxygen carrying capacity whether caused by nutrition, bleeding or high altitude is called
    A) acromegly.
    B) leucopenia.
    C) anemia.
    D) erythroblastosis fetalis.
    E) histoimcompatiblity.
    Answer: C

43. The most abundant blood cell in the body is the
    A) monocytes.
    B) neutrophils.
    C) basophils.
    D) erythrocytes.
    E) eosinophils.
    Answer: D

True-False

Write T if the statement is true and F if the statement is false

1. All blood cells originate from hemopoietic stem cells.
   Answer: True

2. The kidney excretes bilirubin.
   Answer: False

3. Neonatal jaundice disappears as the liver matures.
   Answer: True

4. Leukocytes can be divided into three major groups.
   Answer: False

5. Lymphocytes are the largest leukocytes.
   Answer: False

6. Monocytes that migrate to infected tissues are called wandering macrophages.
   Answer: True

7. Inflammation causes an increase in the hematocrit.
   Answer: False

8. Neutrophils are granulocytes.
   Answer: True

9. Plasma proteins are confined to blood.
   Answer: True

10. The extrinsic pathway of blood clotting is more complex than the intrinsic pathway.
    Answer: False

11. The breaking up of a blood clot is called clot retraction.
    Answer: False

12. Blood clotting always results in the formation of embolus.
    Answer: False

13. Heparin is produced by mast cells.
    Answer: True

14. A blood clot that is transported by the blood stream is called a thrombus.
    Answer: False

15. Agglutinins are also called isoantibodies.
    Answer: True

16. Cyanosis is caused by too many leukocytes in the blood.
    Answer: False

17. The production of red blood cells is closely linked to the carbon dioxide carrying capacity of the blood. If this is low, the red bone marrow makes a hormone called erythroheparin to increase the number of RBC's.
    Answer: False

18. Hematocytoblasts are special cells in the red bone marrow which have the potential to become any of the seven formed elements.
    Answer: True

19. Being type AB means you either have the A or B antigens but not both.
    Answer: False.

20. Plasma constitutes about 55% on the blood volume.
    Answer: True

21. Mature erythrocytes can reproduce.
    Answer: False

Short Answer

Write the word or phrase that best completes each statement or answers the question.

1. Collective erythrocytes, leukocytes, and platelets are called the ____ elements of blood.
   Answer: formed

2. The branch of science concerned with the study of blood is called _________.
   Answer: hematology

3. The three functions of blood are transportation, regulation, and _______.
   Answer: protection

4. Blood belongs to ____ tissue.
   Answer: connective

5. Another name for immunoglobulins is __________.
   Answer: antibodies

6. The process of blood cell formation is called ____.
   Answer: hemopoiesis

7. The pigment that carries oxygen in red blood cells is ____.
   Answer: hemoglobin

8. The protein portion of hemoglobin is _____.
   Answer: globin

9. Erythrocyte formation is called ____.
   Answer: erythropoisis

10. Surface proteins on nucleated cells that can be used to identify tissues for transplantation are _______.
    Answer: major histocompatibility antigens

11. The function of platelets is ____.
    Answer: blood clotting

12. When platelets accumulate and attach to each other they form a mass called the ____.
    Answer: platelet plug

13. Breaking of blood clot is called ____.
    Answer: fibrinolysis

14. The surface of red blood cells contains antigens called ____.
    Answer: isoantigens or agglutinogens

15. Individuals with type A blood have ____ antibodies in the plasma.
    Answer: B

16. The Rh system of blood classification was first discovered in the blood of the ____.
    Answer: Rhesus monkey

17. The hemolysis caused by fetal-maternal incompatibility is called ____.
    Answer: erythroblastosis fetalis or hemolytic disease of the newborn

18. A condition in which the oxygen-carrying capacity to the blood is reduced is called ____.
    Answer: anemia

19. Bleeding, either internal or external is referred to as ____.
    Answer: hemorrhage

20. Septicemia is the medical term for ____.
    Answer: blood poisoning

21. The release of hemoglobin into the blood as a result of red blood cell rupture is called ____.
    Answer: hemolysis

22. During blood clotting prothrombin is converted into the enzyme ____.
    Answer: thrombin

23. Plasma minus the clotting proteins is called ____.
    Answer: serum

Matching

Choose the item from Column 2 that best matches each item in Column 1.

1. Column 1: Contain hemoglobin.
   Column 2: erythrocytes

2. Column 1: Release lysozyme.
   Column 2: neutrophils

3. Column 1: Produce antibodies.
   Column 2: lymphocytes

4. Column 1: Become wandering macrophages.
   Column 2: monocytes

5. Column 1: Release histaminase.
   Column 2: eosinophils

6. Column 1: Present in high numbers during parasitic infection.
   Column 1: eosinophils

7. Column 1: Known as mast cells.
   Column 2: basophils

8. Column 1: Caused by inadequate absorption or excessive loss of iron.
   Column 2: Iron-deficiency anemia

9. Column 1: Due to an insufficient production of erythrocytes.
   Column 2: Pernicious anemia

10. Column 1: An excessive loss of erythrocytes through bleeding.
    Column 2: Hemolytic anemia

11. Column 1: Characterized by distortion in the shape of erythrocytes.
    Column 2: Hemolytic anemia

12. Column 1: A result of the destruction or inhibition of red bond marrow.
    Column 2: Aplastic anemia

13. Column 1: Due to the production of abnormal hemoglobin.
    Column 2: Sickle-cell anemia

Essay

Write your answer in the space provided or on a separate sheet of paper.

1. Describe the functions of blood.
   Answer: Blood transports oxygen, carbon dioxide, nutrients, wastes, and hormones. Blood helps to regulate pH, body temperature, and water content of the cells. Furthermore, it prevents blood loss through clotting and also fights invading microbes.

2. Name the different types of leukocytes.
   Answer: 1. Granulocytes: neutrophils, basophils, and eosinophils.
   2. Agranulocytes: lymphocytes and monocytes.

3. Name the stages of blood clotting.
   Answer: 1. Formation of prothrombinase (prothrombin activator).
   2. Conversion of prothrombin into thrombin.
   3. Conversion of soluble fibrinogen into insoluble fibrin.

CHAPTER 15 The Cardiovascular System: Heart

Multiple-Choice

Choose the one alternative that best completes the statement or answers the question

1. The outer portion of the pericardium is the
   A) epicardium.
   B) serous pericardium.
   C) fibrous pericardium.
   D) parietal pericardium.
   E) endocardium.
   Answer: C

2. The layer of simple squamous epithelium that lines the inside of the myocardium is called
   A) epicardium.
   B) serous pericardium.
   C) fibrous pericardium.
   D) parietal pericardium.
   E) endocardium.
   Answer: E

3. Cardiac muscle fibers are connected with each other by
   A) intercalated discs.
   B) tight junctions.
   C) desmosomes.
   D) neuromuscular junctions.
   E) synapses.
   Answer: A

4. The portion of the heart that consists of cardiac muscle tissue is the
   A) epicardium.
   B) myocardium.
   C) pericardium.
   D) endocardium.
   E) All of the above.
   Answer: B

5. Which portion of the heart has the thickest myocardium?
   A) left atrium
   B) right atrium
   C) left ventricle
   D) right ventricle
   E) They all have equal thickness.
   Answer: C

6. Blood transported by the pulmonary veins returns to the
   A) left atrium.
   B) left ventricle.
   C) right atrium.
   D) right ventricle.
   E) None of the above.
   Answer: A

7. The muscular wall of the left ventricle is thicker than the right wall because it
   A) holds a greater volume of blood.
   B) pumps the blood to the lungs.
   C) pumps the blood to the entire body.
   D) pumps the blood through a smaller valve.
   E) pumps the blood through the heart.
   Answer: C

8. The groove separating the atria from the ventricles is known as
   A) interventricular septum.
   B) coronary sulcus.
   C) interventricular sulcus.
   D) interatrial septum.
   E) septum.
   Answer: B

9. The left ventricle pumps the blood into the
   A) pulmonary trunk.
   B) ascending aorta.
   C) descending aorta.
   D) pulmonary vein.
   E) vena cava.
   Answer: B

10. Which blood vessel delivers blood to the right atrium?
    A) vena cava
    B) pulmonary artery
    C) pulmonary trunk
    D) pulmonary vein
    E) aorta
    Answer: A

11. The blood vessel that carries blood highest in oxygen is the
    A) ascending aorta.
    B) superior vena cava.
    C) pulmonary vein.
    D) coronary artery.
    E) pulmonary trunk.
    Answer: C

12. The valve between the left atrium and left ventricle is the
    A) bicuspid valve.
    B) tricuspid valve.
    C) aortic semilunar valve.
    D) pulmonary semilunar valve.
    E) sinoatrial valve.
    Answer: A

13. The valve between the left ventricle and the blood vessel leaving the left ventricle is the
    A) bicuspid valve.
    B) tricuspid valve.
    C) aortic semilunar valve.
    D) pulmonary semilunar valve.
    E) mitral valve.
    Answer: C

14. The blood vessel that collects deoxygenated blood of the coronary circulation and empties into the right atrium is the
   A) vena cava.
   B) sigmoid sinus.
   C) pulmonary sinus.
   D) coronary sinus.
   E) None of the above.
   Answer: D

15. The part of the circulation pumping blood to and from the lungs is known as
   A) systemic circulation.
   B) coronary circulation.
   C) pulmonary circulation.
   D) respiratory circulation.
   E) hepatic circulation.
   Answer: C

16. Death of an area of cardiac tissue due to an interrupted blood supply is called
   A) ischemia.
   B) myocardial infarction.
   C) angina pectoris.
   D) ventricular fibrillation.
   E) cardiac ischemia.
   Answer: B

17. The blood supply to the myocardial wall of the heart is supplied by the
   A) arch of the aorta.
   B) descending aorta.
   C) pulmonary artery.
   D) coronary sinus.
   E) coronary arteries.
   Answer: E

18. The normal pacemaker of the heart is/are the
   A) Purkinje fibers.
   B) bundle of His.
   C) atrioventricular node.
   D) sinoatrial node.
   E) atrioventricular bundle.
   Answer: D

19. Which of the following represents the correct sequence of structures in the cardiac conduction system?
   A) Purkinje fibers, AV nodes, SA node, bundle of His
   B) AV node, SA node, bundle of His, Purkinje fibers
   C) SA node, bundle of His, AV node, Purkinje fibers
   D) SA node, AV node, Purkinje fibers, bundle of His
   E) SA node, AV node, bundle of His, Purkinje fibers
   Answer: E

20. The P wave of an ECG indicates
   A) atrial depolarization.
   B) atrial repolarization.
   C) ventricular depolarization.
   D) ventricular repolarization.
   E) ventricular contraction.
   Answer: A

21. Atrial repolarization is indicated by the
   A) P wave.
   B) QRS wave.
   C) T wave.
   D) R portion of the QRS.
   E) None of the above.
   Answer: E

22. Which of the following represents the correct order of ECG waves?
   A) QRS, T, P
   B) QRS, P, T
   C) P, QRS, T
   D) T, QRS, P
   E) P, T, QRS
   Answer: C

23. At rest each cardiac cycle lasts approximately _____ seconds.
   A) 0.3
   B) 0.5
   C) 0.8
   D) 0.10
   E) 0.20
   Answer: C

24. The remaining 25% of the blood that fills the ventricles after the relaxation period occurs during
   A) atrial diastole.
   B) atrial systole.
   C) ventricular diastole.
   D) ventricular systole.
   E) restoration period.
   Answer: B

25. Closing of the AV valves produces the
   A) P wave.
   B) T wave.
   C) QRS wave.
   D) the first heart sound (lubb).
   E) the second heart sound (dubb).
   Answer: D

26. Heart murmurs are usually the result of defective
   A) nodes.
   B) valves.
   C) arteries.
   D) myocardial cells.
   E) Purkinje fibers.
   Answer: B

27. The cardiac output is
   A) the total volume of blood within the body.
   B) the milliliters of blood pumped per beat by each ventricle.
   C) the milliliters of blood pumped per minute by each ventricle.
   D) the number of beats per minute.
   E) the total amount of blood pumped by the heart per day.
   Answer: C

28. The cardiovascular center is located in the
    A) hypothalamus.
    B) thalamus.
    C) pons.
    D) midbrain.
    E) medulla oblongata.
    Answer: E

29. Which of the following causes an increase in the rate of the heartbeat?
    A) norepinephrine
    B) acetylcholine
    C) parasympathetic fibers
    D) Both A and C
    E) Both B and C
    Answer: A

30. Changes in the blood pressure are detected by
    A) osmoreceptors.
    B) baroreceptors.
    C) stretch receptors.
    D) chemoreceptors.
    E) proprioceptors.
    Answer: B

31. Impulses carried by means of the vagus nerve are
    A) parasympathetic impulses that increase the heart rate.
    B) parasympathetic impulses that decrease the heart rate.
    C) sympathetic impulses that increase the heart rate.
    D) sympathetic impulses that decrease the heart rate.
    E) autonomic impulses that both increase and decrease the heart rate.
    Answer: B

32. The average heartbeat of a healthy human is
    A) 60 beats per minute.
    B) 65 beats per minute.
    C) 75 beats per minute.
    D) 80 beats per minute.
    E) 85 beats per minute.
    Answer: C

33. The study of the heart and diseases associated with it is known as
    A) pathology.
    B) physiology.
    C) hematology.
    D) oncology.
    E) cardiology.
    Answer: E

34. The bulk of the heart consists of
    A) epicardium.
    B) endocardium.
    C) myocardium.
    D) serous pericardium.
    E) fibrous pericardium.
    Answer: C

35. The two upper chambers of the heart are separated from each other by the
A) interventricular septum.
B) interatrial septum.
C) parietal pericardium.
D) visceral pericardium.
E) coronary sulcus.
Answer: B

36. Which of the following is/are risk factor(s) for the development of heart disease
A) high blood pressure.
B) diabetes mellitus.
C) genetic predisposition.
D) All of the above.
E) None of the above.
Answer: D

37. The valve located between the right atrium and the right ventricle is the
A) tricuspid valve.
B) bicuspid valve.
C) mitral valve.
D) aortic valve.
E) semilunar valve.
Answer: A

38. When blood moves from an atrium to a ventricle
A) the valve is pushed open.
B) the papillary muscles relax.
C) the corordae tendinae slacken.
D) A and B are correct.
E) All of the above.
Answer: E

39. Parasympathetic neurons reach the heart via the ____ cranial nerve.
A) V
B) VI
C) VIII
D) X
E) XII
Answer: D

40. A fluttering of the heart is called
A) cardiac arrest.
B) palpitation.
C) paroxysmal tachycardia.
D) ventricular fibrilation.
E) cardiomegaly.
Answer: B

41. Blood from which three vessels enters the right atrium?
A) the three pulmonary veins
B) the superior and inferior vena cavae and the coronary arteries
C) the superior and inferior vena cavae and the coronary sinus
D) the 2 pulmonary veins and the pulmonary trunk
E) the left and right common carotid and the iliac artery
Answer: C

42. Atherosclerosis is a condition in which
    A) The arteries are "hardened" by calcium deposits in the walls.
    B) Fatty deposits called plaques partially block the arteries.
    C) Vitamin B-12 cannot be absorbed.
    D) The oxygen carrying capacity of the blood is reduced..
    E) The right side pumps harder than the left side.
    Answer: B

43. Why is the right ventricle wall thinner than the left ventricle?
    A) The right ventricle has to pump harder.
    B) The left ventricle has to pump blood further.
    C) If the right pumped as hard as the left, the lungs would burst.
    D) Both B & C
    E) All of the above reasons are true.
    Answer: D

44. Which of the following does not regulate stroke volume?
    A) degree of stretch in the heart
    B) Starling's law
    C) force of contraction
    D) ag.
    E) pressure required and the opening of the valves
    Answer: D

45. Pulmonary edema results from
    A) heart murmurs.
    B) angina pectoris.
    C) ischemia
    D) congestive heart failure.
    E) All of the above.
    Answer: D

True-False

Write T if the statement is true and F if the statement is false.

1. The heart is located in the mediastenum.
   Answer: True

2. All arteries contain oxygenated blood, and all veins contain non-oxygenated blood.
   Answer: False

3. Atrioventricular valves are held closed by chordae tendinae and papillary muscles.
   Answer: True

4. The heart valves prevent the backflow of blood during the cardiac cycle.
   Answer: True

5. The autonomic nervous system initiates the contraction of the heart.
   Answer: False

6. Malfunctioning heart valves is called myocardial infarction.
   Answer: False

7. Blood flow through the heart is caused by changes in the size of the chambers.
   Answer: True

8. The atria receive the blood returning to the heart.
   Answer: True

9. Acetylcholine released by the parasympathetic fibers increases the rate of the heartbeat.
   Answer: False

10. Elevated levels of potassium decrease the heart rate and strength of the contraction.
    Answer: True

11. Mental states such as depression tend to increase the heart rate.
    Answer: False

12. Alcoholism may contribute to the development of heart disease.
    Answer: True

13. The rate of the heartbeat stays constant throughout life.
    Answer: False

14. Exercise increases cardiac efficiency and output.
    Answer: True

15. The pericardium is actually 3 layers. The outer layer is the parietal pericardium that forms the rather inelastic pericardial cavity around the heart. The middle is the visceral pericardium. The inner layer is called either the fibrous pericardium if you are talking about pericardial coverings or epicardium if you are talking about layers of the heart.
    Answer: False

16. Oxygen is sent to the myocardium first through the cardiac arteries.
    Answer: False

17. In general, heart attack risk is low for those who exercise regularly and higher for those who do not.
    Answer: True

18. Any irregular heart rhythm is an arrhythmia.
    Answer: True

19. Ischemia, hypoxia, angina pectoris and myocardial infarction all result from a lack of oxygen to the heart tissue.
    Answer: True

Short Answer

Write the word or phrase that best completes each statement or answers the question.

1. The inner visceral layer of the pericardium is also called the ____.
   Answer: epicardium

2. The scientific study of the normal heart and the diseases associated with it is _______.
   Answer: cardiology

3. The space between the parietal and visceral pericardial membranes is the ____.
   Answer: pericardial cavity

4. Inflammation of the pericardium is called ________.
Answer: pericarditis

5. The thickenings of the sarcolemma that connect the cardiac muscle fibers are ____.
Answer: intercalated discs

6. The backflow of blood through an incompletely closed valve is called ____.
Answer: regurgitation

7. The pouchlike structure on the anterior surface of each atrium is called ________.
Answer: auricle

8. The right ventricle pumps blood into the ____.
Answer: pulmonary trunk

9. The coronary arteries originate as branches of the ____.
Answer: ascending aorta

10. The valve between the right atrium and the right ventricle is the ______ valve.
Answer: tricuspid

11. The large vein in the back of the heart collecting deoxygenated blood is the ____.
Answer: coronary sinus

12. The structure that picks up an action potential from the SA node is/are the ____.
Answer: AV node

13. The medical term for heart attack is _______ _______.
Answer: myocardial infarction

14. A recording of the electrical changes that accompany the heartbeat is called a(n) _____.
Answer: electrocardiogram

15. The spread of the action potential through the ventricles is recorded as the ____ wave of an ECG.
Answer: QRS

16. The specialized tissue capable of conducting action potentials and allowing the heart to beat without direct stimulus from the nervous system is the ____.
Answer: conduction system

17. Contraction of the heart muscle is referred to as ____.
Answer: systole

18. Heart sounds are produced by the ____ of the AV valves and the semilunar valves.
Answer: closing

19. The amount of blood ejected by a ventricle during each contraction is called the ____.
Answer: stroke volume

20. The ____ law of the heart explains the relationship between the stretching of the ventricular wall and the
contraction strength.
Answer: Starling's

21. A general term referring to an irregularity in the rhythm of the heart is ____.
Answer: arrhythmia

22. A procedure that is used to visualize the coronary arteries, chambers, valves, and great vessels is ____.
    Answer: cardiac catheterization

23. The neurotransmitter released by the sympathetic fibers that increase the rate of the heartbeat is ____.
    Answer: norepinephrine

24. An incomplete closure of the interventricular septum causes ____.
    Answer: interventricular septal defect

25. A defect that exists at birth, and usually before, is called a(n) ____.
    Answer: congenital defect

Matching

Choose the item from Column 2 that best matches each item in Column 1.

1. Column 1: The pacemaker of the heart.
   Column 2: SA node

2. Column 1: Located between the right atrium and the right ventricle.
   Column 2: tricuspid valve

3. Column 1: Located between the left atrium and the left ventricle.
   Column 2: bicuspid valve

4. Column 1: Emerge from the bundle branches.
   Column 2: Purkinje fibers

5. Column 1: Consists of cardiac muscle tissue.
   Column 2: myocardium

6. Column 1: The external layer of the heart wall.
   Column 2: epicardium

7. Column 1: Contraction of the heart chambers.
   Column 2: systole

8. Column 1: Relaxation of the heart chambers.
   Column 2: diastole

9. Column 1: Blood vessel associated with the left ventricle.
   Column 2: aorta

10. Column 1: Blood vessel associated with the right ventricle.
    Column 2: pulmonary trunk

Essay

Write your answer in the space provided or on a separate sheet of paper.

1. Briefly describe the diagnostic value of an electrocardiogram.
   Answer: The ECG is used to diagnose abnormal cardiac rhythms and conduction patterns, detect the presence of life, and follow the course of recovery from a heart attack.

2. Name, in correct sequence, the structures involved as the cardiac impulse passes through the heart.
   Answer: SA node, AV node, bundle of His, Purkinje fibers

3. Name the phases of the cardiac cycle.
   Answer: The relaxation period, ventricular filling, and ventricular systole.

4. Define cardiac output.
   Answer: Cardiac output (CO) is the amount of blood ejected by the left ventricle into the aorta per minute. It is calculated as follows: CO = stroke volume x beats per minute.

CHAPTER 16 The Cardiovascular System: Blood Vessels and Circulation

Multiple-Choice

Choose the one alternative that best completes the statement or answers the question

1. The smallest type of blood vessels are
   A) arteries.
   B) arterioles.
   C) venules.
   D) capillaries.
   E) veins.
   Answer: D

2. The endothelium consists of
   A) simple cuboidal epithelium.
   B) simple squamous epithelium.
   C) epithelium and smooth muscle.
   D) smooth muscle only.
   E) epithelium, smooth muscle, and elastic fibers.
   Answer: B

3. The two main methods of capillary exchange are
   A) diffusion and exocytosis.
   B) endocytosis and exocytosis.
   C) diffusion and bulk flow.
   D) diffusion and filtration.
   E) endocytosis and pinocytosis.
   Answer: C

4. Exchange of nutrients and gases between the blood and tissue is the function of the
   A) arteries.
   B) arterioles.
   C) veins.
   D) venules.
   E) capillaries.
   Answer: E

5. Which of the following statement(s) is/are true for bulk flow?
   A) It is a passive transport mechanism.
   B) It is an active transport mechanism.
   C) It moves large numbers of molecules.
   D) A and C
   E) All of the above.
   Answer: D

6. Venous return is due to
   A) contraction of the heart.
   B) skeletal muscle pump.
   C) respiratory pump.
   D) A only.
   E) A, B, and C.
   Answer: E

7. Blood vessels that are composed of a single layer of endothelial cells and a basement membrane are
   A) venules.
   B) arterioles.
   C) capillaries.
   D) Both A and B.
   E) None of the above.
   Answer: C

8. Venules
   A) are small veins.
   B) collect blood from capillaries.
   C) drain into veins.
   D) are similar in structure to arterioles.
   E) All of the above.
   Answer: E

9. Which of the following blood vessels are referred to as blood reservoirs?
   A) arteries
   B) veins
   C) capillaries
   D) All of the above.
   E) None of the above.
   Answer: B

10. Blood pressure is highest in the
   A) arteries.
   B) arterioles.
   C) veins.
   D) venules.
   E) capillaries.
   Answer: A

11. Resistance is related to
   A) blood viscosity.
   B) blood vessel length.
   C) blood vessel radius.
   D) All of the above.
   E) A and B only.
   Answer: D

12. Which of the following are involved in the regulation of blood pressure?
   A) baroreceptor reflexes
   B) chemoreceptor reflexes
   C) carotid bodies
   D) A and B
   E) A, B, and C
   Answer: E

13. Atrial natriuretic peptide
   A) is released by cells of the heart.
   B) lowers blood pressure.
   C) increases blood pressure.
   D) A and B.
   E) B and C.
   Answer: D

14. Which of the following is NOT a symptom of shock?
    A) decreased epinephrine levels
    B) increased levels of aldosterone
    C) rapid, resting heart rate
    D) cool, pale skin
    E) sweating
    Answer: A

15. Which of the following increase(s) blood pressure?
    A) increased cardiac rate
    B) increased peripheral resistance
    C) increased blood volume
    D) increased water retention
    E) All of the above.
    Answer: E

16. Peripheral resistance
    A) increases as blood viscosity increases.
    B) decreases as blood viscosity increases.
    C) increases with increased blood vessel diameter.
    D) is higher in arteries than capillaries.
    E) decreases with increasing blood vessel length.
    Answer: A

17. The cardiovascular center is located in the
    A) hypothalamus.
    B) thalamus.
    C) medulla oblongata.
    D) midbrain.
    E) cerebral cortex.
    Answer: C

18. Sympathetic stimulation results in
    A) vasodilation.
    B) decrease in blood pressure.
    C) increase in blood pressure.
    D) A and B are correct.
    E) A and C are correct.
    Answer: C

19. Baroreceptors are located in the
    A) aorta.
    B) internal carotid arteries.
    C) jugular veins.
    D) In both A and B.
    E) In both B and C.
    Answer: D

20. Neurons that monitor carbon dioxide levels in the blood and are located in the carotid and aortic bodies are
    A) baroreceptors.
    B) chemoreceptors.
    C) mechanoreceptors.
    D) sympathetic neurons.
    E) parasympathetic neurons.
    Answer: B

21. Which of the following hormones influences blood pressure?
   A) ADH
   B) epinephrine
   C) norepinephrine
   D) renin
   E) All of these hormones influence blood pressure
   Answer: E

22. All of the following hormones cause an increase in blood pressure EXCEPT
   A) epinephrine.
   B) norepinephrine.
   C) renin.
   D) ANP.
   E) ADH.
   Answer: D

23. Which of the following blood vessels would have the greatest resistance to blood flow?
   A) a long, large diameter blood vessel
   B) a short, large diameter blood vessel
   C) a long, small diameter blood vessel
   D) a short, small diameter blood vessel
   E) a medium length, large diameter blood vessel
   Answer: C

24. The movement of blood from the abdominal veins into the thoracic veins is mostly due to
   A) pumping action of the heart.
   B) skeletal muscle contraction.
   C) pressure difference between the abdominal and thoracic cavity.
   D) the increasing blood pressure in the abdominal veins.
   E) All of the above.
   Answer: C

25. All of the following blood vessels are commonly used to feel the pulse EXCEPT
   A) radial artery.
   B) brachial artery.
   C) popliteal artery.
   D) femoral artery.
   E) common carotid artery.
   Answer: D

26. The pulse is a direct reflection of the
   A) cardiac output.
   B) blood pressure.
   C) venous return.
   D) cardiac rate.
   E) None of the above.
   Answer: D

27. The diastolic blood pressure
   A) provides information about the resistance of blood vessels.
   B) indicates the force of ventricular contraction.
   C) indicates the force of atrial contraction.
   D) is normally higher than the systolic blood pressure.
   E) indicates the cardiac output.
   Answer: A

28. Which of the following describes the pulmonary circulation?
A) right atrium to left ventricle
B) left ventricle to right atrium
C) right ventricle to left atrium
D) left atrium to right ventricle
E) left ventricle to left atrium
Answer: C

29. The blood vessel(s) containing blood with the highest oxygen contents is/are the
A) pulmonary arteries.
B) pulmonary veins.
C) ascending aorta.
D) descending aorta.
E) thoracic aorta.
Answer: B

30. All systemic blood vessels branch from the
A) vena cava.
B) pulmonary trunk.
C) circle of Willis.
D) aorta.
E) None of the above.
Answer: D

31. The circulatory route which brings blood to the tissues and back to the heart is the
A) hepatic portal circulation.
B) pulmonary circulation.
C) aortic circulation.
D) systemic circulation.
E) cerebral circulation.
Answer: D

32. The circle of Willis is part of the
A) pulmonary circulation.
B) cerebral circulation.
C) hepatic portal circulation.
D) systemic circulation.
E) common circulation.
Answer: B

33. The blood rich in substances absorbed from the gastrointestinal tract is carried by the
A) fetal circulation.
B) cerebral circulation.
C) hepatic circulation.
D) systemic circulation.
E) pulmonary circulation.
Answer: C

34. Which of the following are changes in the cardiovascular system related to the aging process?
A) reduced cardiac output
B) increased systolic blood pressure
C) decreased maximum heart rate
D) A and B
E) A, B, and C
Answer: E

35. The fetal circulation differs from the adult circulation because of two organ systems that are non-functional. These organ systems are the:
A) cardiovascular and respiratory systems.
B) cardiovascular and digestive systems.
C) respiratory and digestive systems.
D) nervous and respiratory systems.
E) nervous and digestive systems.
Answer: C

36. The coronary arteries branch from the
A) arch of the aorta.
B) thoracic artery.
C) descending aorta.
D) ascending aorta.
E) None of the above.
Answer: D

37. Which of the following arteries does NOT directly branch from the abdominal aorta?
A) internal iliac
B) common iliac
C) celiac trunk
D) superior mesenteric
E) renal
Answer: A

38. The direct continuation of the brachial artery is the ________ artery.
A) ulnar
B) axillary
C) brachiocephalic
D) radial
E) digital
Answer: D

39. The main vein draining blood from the heart tissue is the
A) superior vena cava.
B) brachiocephalic vein.
C) cardiac vein.
D) coronary sulcus.
E) coronary sinus.
Answer: E

40. The external jugular veins empty into the
A) dural venous sinuses.
B) internal jugular veins.
C) subclavian veins.
D) brachiocephalic veins.
E) superior vena cava.
Answer: C

41. The formation of new blood vessels is referred to as
A) syncope.
B) angiogenesis.
C) occlusion.
D) aortography.
E) osteogenesis.
Answer: B

42. Where is blood flow the slowest?
A) arteries
B) arterioles
C) capillaries
D) veins
E) heart
Answer: Veins

43. Which of the following is NOT true
A) Arteries are stronger than veins.
B) Arteries are larger in diameter than veins.
C) Veins have valves, most arteries do not.
D) Arteries have higher blood pressure than veins.
E) All of the above are true statements.
Answer: B

44. Blood flow through a particular tissue area is controlled by
A) constriction of the venous walls.
B) constriction of the precapillary sphincter of the venules.
C) constriction of the precapillary sphincter of the arterioles.
D how fast the heart is beating.
E) the pulse rate.
Answer: C

45. How is blood mainly moved through the veins?
A) by the pressure supplied by the heart beat
B) by changes in local muscle tissue and thoracic pressure.
C) by peristalsis of the muscular tissue of the veins.
D) by the beating of in the vein's endothelium.
E) both A & C.
Answer: B

46. What vessel sends blood to the heart myocardium from the aorta?
A) superior vena cava
B) coronary arteries
C) cardiac arteries
D) coronary sinus
E) Purkinje fibers
Answer: B

47. Which of the following represents the correct sequence of parts through which blood moves in passing from the superior and inferior vena cavae to the lungs?
A) right atrium, pulmonary artery, right ventricle, pulmonary vein
B) right atrium, right ventricle, pulmonary trunk
C) right atrium, right ventricle, pulmonary veins
D) left atrium, left ventricle, pulmonary trunk
E) left atrium, left ventricle, ascending aorta, aortic arch
Answer: B

48. Gases and nutrients leave the blood at the
    A) veins.
    B) arterioles.
    C) venous end of the capillaries.
    D) arteriole end of the capillaries.
    E) venules.
    Answer: D

49. Blood flow from the heart to the lungs is called
    A) the pulmonary circuit.
    B) the circuit court.
    C) the Indy circuit.
    D) the electrical circuit.
    E) the systemic circuit.
    Answer: A

50. Atherosclerosis is a condition in which
    A) the arteries are "hardened" by deposits within the wall.
    B) fatty deposits in the lumen partially blocking the arteries.
    C) Vitamin B12 cannot be absorbed.
    D) the oxygen carrying capacity of the blood is reduced.
    E) the immune system is compromised.
    Answer: A

True-False

Write T if the statement is true and F if the statement is false.

1. Sympathetic stimulation causes vasoconstriction.
   Answer: True

2. All veins carry blood low in oxygen.
   Answer: False

3. Leaky valves cause varicose veins.
   Answer: True

4. Blood flow refers to the amount of blood that passes through a blood vessel in a given period of time.
   Answer: True

5. Blood always flows from regions of lower blood concentration to regions of higher blood concentration.
   Answer: False

6. Epinephrine has no effect on blood pressure.
   Answer: False

7. Peripheral resistance refers to the resistance to blood flow in the peripheral circulation.
   Answer: True

8. Blood volume only occasionally changes blood pressure.
   Answer: False

9. The vasomotor center is located in the hypothalamus.
Answer: False

10. Parasympathetic stimulation via the vagus nerve causes a decrease in heart rate.
Answer: True

11. Depression can cause a decrease in blood pressure.
Answer: True

12. Muscle tissue is not dependent on autoregulation of blood flow.
Answer: False

13. Velocity of blood flow depends on the cross-sectional area of the blood vessel.
Answer: True

14. Skeletal muscle contractions help to return venous blood.
Answer: True

15. Bradycardia indicates a rapid heart rate.
Answer: False

16. Arteries and veins have several layers including two layers of connective tissue, a layer of smooth muscle and a smooth endothelium lining for efficient blood flow and to prevent clotting. The muscle lining is thicker in the arteries than it is in the veins.
Answer: True

17. There isn't enough blood to go around to all the tissues so the diameter of the arteries and arterioles as well as the constriction of the precapillary sphincters helps determine where blood should go.
Answer: True

18. Heart rate is regulated by the cardiac center in the hypothalamus and chemoreceptors that detect oxygen levels in the blood.
Answer: False

19. The major factor that moves blood through the body is pressure differences in the blood vessels between when it leaves the heart and as it returns.
Answer: True

20. Systolic pressure is the higher number because it is the pressure in the arteries when the ventricles are contracting. Diastolic pressure is lower but it is a constant pressure, because the ventricles are relaxed.
Answer: True

Short Answer

Write the word or phrase that best completes each statement or answers the question.

1. Arterioles within a tissue or organ branch into countless microscopic vessels called ____.
Answer: capillaries

2. The movement of water and solutes out of capillaries into the interstitial fluid is called _____.
Answer: filtration

3. The movement of water and solutes from the interstitial fluid into the capillaries is called _____.
Answer: reabsorption

4. An increase in the size of a blood vessel lumen is referred to as ____.
Answer: vasodilatation

5. The blood flow in capillaries is regulated by smooth muscle fibers called the ____.
Answer: pericapillary sphincter

6. Chemoreceptors that are sensitive to carbon dioxide levels are found in the carotid and ____ bodies.
Answer: aortic

7. Veins that lose their elasticity and become stretched and flabby are called ______ veins.
Answer: varicose

8. Chemicals that alter the diameter of blood vessels are referred to as ____ factors.
Answer: vasoactive

9. When hydrostatic pressure in capillaries is higher than osmotic pressure, ____ occurs.
Answer: filtration

10. As blood flows from capillaries to venules to veins to the heart, its velocity ____.
Answer: increases

11. The alternate expansion and recoiling of an artery with each contraction of the left ventricle is called ____.
Answer: pulse

12. The instrument used to measure blood pressure is called ____.
Answer: sphygmomanometer

13. Sweating during shock is due to ____ stimulation.
Answer: sympathetic

14. The section between the diaphragm and the common iliac arteries is referred to as the ____.
Answer: abdominal aorta

15. The blood vessel emerging from the right ventricle is the ____.
Answer: pulmonary trunk

16. A dislodged thrombus which found its way into the pulmonary arterial blood flow causes ____.
Answer: pulmonary embolism

17. A small artery is called a(n) ____.
Answer: arteriole

18. Cardiac output is calculated by multiplying stroke volume by ____.
Answer: heart rate

19. Blood flow from the left ventricle to the right atrium is the ____ circulation.
Answer: systemic

20. The abdominal aorta branches into the ____.
Answer: common iliac arteries

21. The cerebral circulation is a subdivision of the ____ circulation.
    Answer: systemic

22. All veins of the systemic circulation flow into either the superior or inferior vena cava or the ____.
    Answer: coronary sinus

23. The two pumps which help to return the blood to the heart are the skeletal muscle pump and the ____ pump.
    Answer: respiratory

24. Sympathetic impulses reach the heart via ______ ______ nerves.
    Answer: cardiac accelerator

25. The hormone released by cells of the kidneys in response to decreased blood volume or flow is _____.

    Answer: renin

Matching

Choose the item from Column 2 that best matches each item in Column 1.

1. Column 1: Refers to rapid heart rate.
   Column 2: tachycardia

2. Column 1: Indicates a slow heart rate.
   Column 2: bradycardia

3. Column 1: Low blood pressure
   Column 2: hypotension

4. Column 1: High blood pressure
   Column 2: hypertension

5. Column 1: Inflammation of an artery.
   Column 2: arteritis

6. Column 1: Inflammation of a vein.
   Column 2: phlebitis

7. Column 1: Inflammation of a vein with clot formation.
   Column 2: thrombophlebitis

8. Column 1: The obstruction of a blood vessel lumen.
   Column 2: occlusion

9. Column 1: A temporary cessation of consciousness.
   Column 2: syncope

Essay

Write your answer in the space provided or on a separate sheet of paper.

1. Describe the factors that affect blood pressure.

   Answer:
   1. Any factor that increases cardiac output increases blood pressure.
   2. An increase in blood volume also increases blood pressure.
   3. Peripheral resistance to blood flow influences blood pressure. Arterioles control peripheral resistance – and therefore blood pressure – by changing their diameters.

2. Name the factors that help venous return to the heart.

   Answer:
   1. Pumping action of the heart.
   2. Velocity of blood flow.
   3. Skeletal muscle contractions.
   4. Valves in the veins.
   5. Breathing.

3. Briefly describe the systemic circulation.

   Answer: Left ventricle – ascending aorta – arch of the aorta – descending aorta – thoracic aorta – abdominal aorta – organs and tissues – inferior and superior vena cava – right atrium.

CHAPTER 17 The Lymphatic and Immune System, and Resistance to Disease

Multiple-Choice

Choose the one alternative that best completes the statement or answers the question

1. Lymphatic tissue is a specialized form of
   A) loose connective tissue.
   B) reticular connective tissue.
   C) elastic connective tissue.
   D) epithelial tissue.
   E) glandular tissue.
   Answer: B

2. All of the following belong to the lymphatic system EXCEPT
   A) lymph.
   B) lymphatic vessels.
   C) red bone marrow.
   D) yellow bone marrow.
   E) All of the above.
   Answer: D

3. Which of the following vitamins are transported by the lymphatic system?
   A) Vitamin A
   B) Vitamin E
   C) Vitamin C
   D) Both A and E
   E) Both B and C
   Answer: D

4. The thoracic duct
   A) receives lymph from the left side of the head.
   B) receives lymph from the upper right side of the body.
   C) empties the lymph into the right subclavian vein.
   D) both B and C are correct.
   E) both A and C are correct.
   Answer: A

5. Which of the following cells produce antibodies?
   A) T-lymphocytes
   B) B-lymphocytes
   C) monocytes
   D) phagocytes
   E) erythrocytes
   Answer: B

6. Which of the following is NOT surrounded by a connective tissue capsule?
   A) spleen
   B) thymus gland
   C) lymph nodes
   D) lymphatic nodules
   E) All of the above are surround by a capsule
   Answer: D

7. The lymphatic system
   A) transports lipids from the gastrointestinal tract to the blood.
   B) returns filtered proteins to the cardiovascular system.
   C) returns excess interstitial fluid to the cardiovascular system.
   D) functions in surveillance and defense.
   E) All of the above are correct.
   Answer: E

8. Lymphatic vessels
   A) resemble arteries.
   B) resemble arterioles.
   C) resemble veins.
   D) have less valves than veins.
   E) return lymph directly to the heart.
   Answer: C

9. Which of the following is the largest single mass or lymphatic tissue?
   A) tonsils
   B) spleen
   C) lymph nodes
   D) bone marrow
   E) lymphatic nodules
   Answer: B

10. Edema can be a result of
    A) an infected lymph node.
    B) blockage of a lymphatic vessel.
    C) increased blood pressure.
    D) A and B are correct.
    E) A, B, and C are correct.
    Answer: E

11. Lymph nodes
    A) are bean-shaped organs.
    B) are located along lymphatic vessels.
    C) filter the lymph.
    D) are scattered throughout the body.
    E) All of the above.
    Answer: E

12. Afferent lymphatic vessels are characteristic for
    A) tonsils.
    B) the spleen.
    C) lymphatic nodules.
    D) lymph nodes.
    E) the thymus gland.
    Answer: D

13. Worn-out and damaged red blood cells are destroyed in the
    A) thymus gland.
    B) tonsils.
    C) spleen.
    D) lymph nodes.
    E) lymphatic vessels.
    Answer: C

14. Hemopoiesis during fetal life occurs in the
   A) liver.
   B) spleen.
   C) thymus gland.
   D) lymph nodes.
   E) None of the above are correct.
   Answer: B

15. T-cell maturation is a function of the
   A) liver.
   B) spleen.
   C) tonsils.
   D) thymus.
   E) All of the above.
   Answer: D

16. Which of the following is/are characteristic structure(s) of the spleen?
   A) efferent lymphatic vessels
   B) afferent lymphatic vessels
   C) T-lymphocytes
   D) red pulp
   E) Both A and B.
   Answer: D

17. Which of the following carry out the function of the spleen after splenectomy?
   A) thymus
   B) lymph nodes
   C) liver
   D) red bone marrow
   E) both C and D
   Answer: E

18. The first line of defense against pathogens is
   A) phagocytosis.
   B) production of antibodies.
   C) inflammation.
   D) the intact skin.
   E) immunity.
   Answer: D

19. All of the following are characteristic for the second line of defense EXCEPT
   A) phagocytes.
   B) sebum.
   C) antimicrobial proteins.
   D) inflammation.
   E) fever.
   Answer: B

20. Which of the following is considered a chemical factor in the first line of defense?
   A) flow of urine
   B) mucous membrane
   C) saliva
   D) vomiting
   E) gastric juice
   Answer: E

21. Complement proteins
    A) are found in blood plasma.
    B) are present in infected cells.
    C) are produced by T-cells.
    D) are interferons.
    E) are produced by B-lymphocytes.
    Answer: A

22. Macrophages develop from
    A) lymphocytes.
    B) monocytes.
    C) neutrophils.
    D) basophils.
    E) eosinophils.
    Answer: B

23. Which of the following organs does NOT contain fixed macrophages?
    A) lungs
    B) brain
    C) liver
    D) spleen
    E) All of the above contain fixed macrophages.
    Answer: E

24. All of the following are symptoms of inflammation EXCEPT
    A) pain.
    B) redness.
    C) fever.
    D) swelling.
    E) heat.
    Answer: C

25. Which of the following can act as an antigen?
    A) bacteria
    B) viruses
    C) pollen
    D) food
    E) All of the above.
    Answer: E

26. Vaccination is an example of
    A) naturally acquired active immunity.
    B) naturally acquired passive immunity.
    C) artificially acquired active immunity.
    D) artificially acquired passive immunity.
    E) None of the above.
    Answer: C

27. Mother's milk will provide the infant with
    A) naturally acquired active immunity.
    B) naturally acquired passive immunity.
    C) artificially acquired active immunity.
    D) artificially acquired passive immunity.
    E) lifelong immunity.
    Answer: B

28. Cell mediated immunity is provided by
    A) phagocytes.
    B) macrophages.
    C) basophils.
    D) T-cells.
    E) B-cells.
    Answer: D

29. Which of the following stimulate the development of B-cells into antibody-producing plasma cells?
    A) suppressor T-cells
    B) cytotoxic T-cells
    C) helper T-cells
    D) memory T-cells
    E) delayed hypersensitivity cells
    Answer: C

30. The cells that are important in preventing autoimmune disease are
    A) memory T-cells.
    B) helper T-cells.
    C) plasma cells.
    D) suppressor T-cells.
    E) memory B-cells.
    Answer: D

31. The first antibody present at an initial contact with an antigen is
    A) IgA.
    B) IgG.
    C) IgE.
    D) IgD.
    E) IgM.
    Answer: E

32. A cell infected by a virus will produce
    A) antibodies.
    B) immunoglobulins.
    C) prostaglandins.
    D) interferons.
    E) interleukin.
    Answer: D

33. The initial response to HIV invasion is a modest decline in the number of circulating
    A) monocytes.
    B) B-lymphocytes.
    C) cytotoxic T-cells.
    D) memory T-cells.
    E) helper T-cells.
    Answer: E

34. Lymph nodes are important because they
    A) filter the blood.
    B) program T-cells to recognize antigens.
    C) filter lymph and store lymphocytes.
    D) pump lymph through the vessels back to the heart.
    E) All of the above.
    Answer: C

35. The spleen is unique among the lymphatic organs because it stores
   A) T-cells.
   B) B-cells.
   C) blood.
   D) antibodies.
   E) memory cells.
   Answer: C

36. T cells received their name because they
   A) are made in the thyroid.
   B) are produced under the influence of testosterone.
   C) are controlled by the thalamus.
   D) mature in the thymus.
   E) come after S cells.
   Answer: D

37. A plasma cell
   A) secretes antibodies.
   B) starts as a T-cell until it is exposed to its specific antigen.
   C) starts as a B- cell until it is exposed to its specific antigen.
   D) A and B
   E) A & C
   Answer E

38. Which of the following is an example of passive, naturally acquired immunity?
   A) Mom passes her antibodies on through her breast milk.
   B) Dr. Seuss gave you a rabies serum.
   C) The good doctor gave you the polio vaccine.
   D) Your silly cousin exposed you to a nasty cold just before spring break.
   E) None of the above are passive and naturally acquired.
   Answer: A

39. Why is inflammation beneficial for an injury site?
   A) It increases the number of phagocytes in the area.
   B) It walls off the infection to prevent it from spreading.
   C) In increases the temperature of the area.
   D) Increased blood flow to the area helps remove dead cells and toxins.
   E) All of the above are benefits of inflammation.
   Answer: E

True-False

Write T if the statement is true and F if the statement is false.

1. When fluid bathes the cells, it is called interstitial fluid, when it flows through lymphatic vessels it is called lymph.
   Answer: True

2. The thymus gland is located in the thoracic cavity.
   Answer: True

3. Edema is an accumulation of blood in tissue spaces.
   Answer: False

4. Immunity is the production of specific antibodies against a specific pathogen or toxin.
   Answer: True

5. Body cells infected with bacteria produce proteins called interferons.
   Answer: False

6. Phagocytosis is the ingestion of microbes or any foreign particles or cell debris by phagocytes.
   Answer: True

7. Macrophages that remain in certain tissues and organs of the body are called fixed macrophages.
   Answer: True

8. Only severe inflammations produce pus.
   Answer: False

9. Only microbes are considered antigens in the human body.
   Answer: False

10. Before T-cells can attack antigens, the T-cells must become sensitized.
    Answer: True

11. Cytotoxic T-cells destroy target cells on contact.
    Answer: True

12. Delayed hypersensitivity cells are activated in response to allergic reactions.
    Answer: True

13. Memory B-cells produce antibodies.
    Answer: False

14 It has been shown that mosquitoes can transmit the HIV virus.
   Answer: False

15. Lymph is similar to interstitial fluid in composition because lymph is tissue fluid that has moved into lymph capillaries.
    Answer: True

16. The main difference between antibody mediated immunity and cell mediated immunity is that the first one attacks the pathogens directly (B-cells) and the latter produces chemicals to attack them (T-cells).
    Answer: False

17. Fat soluble nutrients are transported in the lymphatic system.
    Answer: True

18. There are five larger clusters of lymphatic nodules called the tonsils. They include the pharyngeal tonsil, the paired palatine and lingual tonsils.
    Answer: True

19. The complement system is a series of proteins that help other defense systems fight pathogens by harming their plasma cell membrane.
    Answer: True

Short Answer

Write the word or phrase that best completes each statement or answers the question.

1. Lymphatic vessels lymph to the venous blood in the ________ veins.
   Answer: subclavian

2. Lymph leaves a lymph node through the ____ lymphatic vessel.
   Answer: efferent

3. The main collecting duct of the lymphatic system that receives lymph from the left side of the body is the ____ _____.
   Answer: thoracic duct

4. The "milking action" of skeletal muscle contractions is referred to as the _____ _____ _____.
   Answer: skeletal muscle pump

5. The spread of cancer cells from a primary tumor to other sites of the body with the help of the lymphatic system is referred to as ________.
   Answer: metastasis

6. Disease producing organisms are called ____.
   Answer: pathogens

7. The ability to ward off disease is called ____.
   Answer: resistance

8. Defecation and vomiting are examples of the _____ line of defense.
   Answer: first

9. The lack of resistance is called ____.
   Answer: susceptibility

10. Some complement proteins produce holes in the plasma membrane of microbes causing the contents of the microbes to leak out by a process called ____.
    Answer: cytolysis

11. When an infection occurs, neutrophils and ____ migrate to the infected area.
    Answer: monocytes

12. Materials indigestible by phagocytes remain in cellular structures called ____.
    Answer: residual bodies

13. The injection of antibodies from an outside source is an example of ____ immunity.
    Answer: artificially acquired passive

14. A protein produced by plasma cells in response to the presence of a foreign substance is called ____.
    Answer: antibody

15. Cells believed to depress parts of the immune response are called ____.
    Answer: suppressor T-cells

16. During a secondary immune response plasma cells produce mainly ____ antibodies.
    Answer: IgG

17. Despite immunologic surveillance, some cancer cells escape destruction, a phenomenon called ____.
Answer: immunologic escape

18. Infections that take advantage of a person's compromised immune system are called ____ infections.
Answer: opportunistic

19. A person who is overly reactive to an antigen is said to be ____.
Answer: hypersensitive (allergic)

20. The antigens that induce an allergic reaction are called ____.
Answer: allergens

21. A tumor composed of lymphatic tissue is called ____.
Answer: lymphoma

22. The replacement of an injured or diseased tissue or organ is called ____.
Answer: transplantation

23. Iron-binding proteins that inhibit the growth of certain bacteria are called ______.
Answer: transferrins

Matching

Choose the item from Column 2 that best matches each item in Column 1.

1. Column 1: A form of cancer usually arising in lymph nodes.
Column 2: Hodgkin's disease

2. Column 1: An autoimmune, noncontagious, inflammatory disease of connective tissue.
Column 2: Systemic lupus erythematosus

3. Column 1: A condition suffered by the "bubble boy".
Column 2: Severe combined immunodeficiency

4. Column 1: Enlarged, tender, and inflamed lymph nodes.
Column 2: Adenitis

5. Column 1: Inflammation of lymphatic vessels.
Column 2: Lymphangitis

6. Column 1: A benign tumor of the lymphatic vessels.
Column 2: lymphangioma

7. Column 1: Ingestion and destruction of microbes.
Column 2: phagocytosis

8. Column 1: Engulfing a microbe into a phagocytotic vesicle.
Column 2: ingestion

9. Column 1: Attachment of macrophage to a microbe.
Column 2: adherence

10. Column 1: Promotion of phagocytosis by coating with a complement.
Column 2: opsonization

Essay

Write your answer in the space provided or on a separate sheet of paper.

1. Name the functions of the lymphatic system.

   Answer: 1. Draining intestinal fluid.
   2. Transporting lipids and lipid-soluble vitamins from the gastrointestinal tract to the blood.
   3. Protection of the body from foreign cells, microbes, and cancer cells.

2. List the mechanical factors in the first line of defense against disease causing microorganisms.

   Answer: skin, mucous membranes, lacrimal apparatus, saliva, mucus, hairs, cilia, epiglottis, flow of urine, defecation, vomiting.

3. Describe the phases of phagocytosis.

   Answer: 1. Chemotaxis is the chemical attraction of phagocytes to microorganisms. These chemicals might come from the invading microbes, white blood cells. Damaged tissue cells, or activated complement.
   2. Adherence, the phagocyte attaches itself to the microorganism or other foreign material.
   3. Ingestion, the phagocyte engulfs the microorganism or foreign substance with the help of pseudopods.

CHAPTER 18 The Respiratory System

Multiple-Choice

Choose the one alternative that best completes the statement or answers the question.

1. Which of the following belongs to the respiratory portion of the lower respiratory system?
   A) larynx
   B) trachea
   C) terminal bronchioles
   D) alveolar ducts
   E) both C and D
   Answer: D

2. The exchange of gases between blood and cells is called
   A) pulmonary ventilation.
   B) internal respiration.
   C) external respiration.
   D) cellular respiration.
   E) inspiration.
   Answer: B

3. Which of the following does NOT belong to the conducting portion of the respiratory system?
   A) alveolar ducts
   B) terminal bronchioles
   C) bronchioles
   D) nose
   E) pharynx
   Answer: A

4. The internal part of the nose is connected to the pharynx through the
   A) nasal septum.
   B) nasal conchae.
   C) paranasal sinuses.
   D) external nares.
   E) internal nares.
   Answer: E

5. The nasal cavity is lined with
   A) simple columnar epithelium.
   B) transitional epithelium.
   C) pseudostratified ciliated columnar epithelium.
   D) stratified ciliated columnar epithelium.
   E) simple ciliated cuboidal epithelium.
   Answer: C

6. The palatine tonsils are found in the
   A) nasopharynx.
   B) oropharynx.
   C) laryngopharynx.
   D) larynx.
   E) None of the above.
   Answer: B

7. The structure with openings to the Eustachian tubes is the
   A) larynx.
   B) fauces.
   C) oropharynx.
   D) nasopharynx.
   E) laryngopharynx.
   Answer: D

8. The nasal cavity is divided into right and left sides by the
   A) external nares.
   B) internal nares.
   C) nasal conchae.
   D) nasal septum.
   E) paranasal sinuses.
   Answer: D

9. Which of the following is called the Adam's apple?
   A) thyroid cartilage
   B) epiglottis
   C) cricoid cartilage
   D) arytenoids cartilage
   E) glottis
   Answer: A

10. The structure which closes off the larynx is the
    A) glottis.
    B) Adam's apple.
    C) epiglottis.
    D) tongue.
    E) vocal cords.
    Answer: C

11. The greater the pressure of air against the vocal cords,
    A) the higher the pitch.
    B) the lower the sound.
    C) the louder the sound.
    D) the lower the pitch.
    E) the hoarser the sound.
    Answer: C

12. The false vocal cords
    A) aid in the production of sound.
    B) are located in the lower larynx.
    C) produce high pitch sounds.
    D) All of the above.
    E) None of the above.
    Answer: E

13. Which of the following describes a correct order of structures in the respiratory passageways?
    A) pharynx, trachea, larynx, bronchi, bronchioles
    B) larynx, pharynx, trachea, bronchioles, bronchi
    C) trachea, pharynx, larynx, bronchi, bronchioles
    D) pharynx, larynx, trachea, bronchi, bronchioles
    E) pharynx, larynx, trachea, bronchioles, bronchi
    Answer: D

14. The C-shaped rings that provide support for the wall of the trachea are made out of
   A) hyaline cartilage.
   B) elastic cartilage.
   C) fibrocartilage.
   D) reticular connective tissue.
   E) bone.
   Answer: A

15. Histamine
   A) has no effect on bronchioles.
   B) causes bronchiole constriction.
   C) causes bronchiole dilation.
   D) causes muscle spasm.
   E) causes inflammation of the bronchioles.
   Answer: B

16. Epinephrine
   A) has no effect on bronchioles.
   B) causes bronchiole constriction.
   C) causes bronchiole dilation.
   D) causes muscle spasm.
   E) causes inflammation of the bronchioles.
   Answer: C

17. The pain of pleurisy is caused by
   A) bronchiole constriction.
   B) bronchiole dilation.
   C) friction between the swollen membranes.
   D) histamine release.
   E) parasympathetic stimulation.
   Answer: C

18. Which of the following is NOT a structure associated with the lungs?
   A) visceral pleura
   B) parietal pleura
   C) cardiac notch
   D) endocardium
   E) hilus
   Answer: D

19. The exchange of gases occurs in the
   A) trachea.
   B) terminal bronchioles.
   C) alveoli.
   D) primary bronchus.
   E) All of the above.
   Answer: C

20. Surfactant is produced by
   A) terminal bronchioles.
   B) alveolar macrophages.
   C) squamous pulmonary epithelial cells.
   D) Surfactant-secreting cells.
   E) capillary membranes.
   Answer: D

21. Which of the following divides into alveolar ducts?
    A) secondary bronchioles
    B) respiratory bronchioles
    C) tertiary bronchioles
    D) terminal bronchioles
    E) alveoli
    Answer: B

22. For air to enter the lungs during inspiration
    A) the pressure inside the lungs must become lower than the atmospheric pressure.
    B) the pressure inside the lungs must be higher than the atmospheric pressure.
    C) the pressure inside the lungs must be equal to the atmospheric pressure.
    D) the size of the lungs must be decreased.
    E) the diaphragm has to be relaxed.
    Answer: A

23. The lungs contain about _______ alveoli.
    A) 10,000
    B) 300,000
    C) one million
    D) 300 million
    E) 500 million
    Answer: D

24. The volume of air that can be exhaled after normal exhalation is the
    A) tidal volume.
    B) residual volume.
    C) inspiratory reserve volume.
    D) expiratory reserve volume.
    E) minute volume of respiration.
    Answer: D

25. The volume of air in a normal breath is called
    A) total lung capacity.
    B) vital capacity.
    C) tidal volume.
    D) functional residual capacity.
    E) residual volume.
    Answer: C

26. The total lung capacity equals
    A) TV + IRV + ERV + RV.
    B) IRV + ERV + RV.
    C) TV + IRV.
    D) FRC + ERV + RV.
    E) TV + ERV + IRV.
    Answer: A

27. Gas exchange in the lungs happens by the process of
    A) osmosis.
    B) diffusion.
    C) exocytosis.
    D) active transport.
    E) filtration.
    Answer: B

28. All of the following decrease the efficiency of external respiration EXCEPT
A) emphysema.
B) increased alveolar PO_2.
C) increased altitude.
D) pneumonia.
E) high alveolar PCO_2.
Answer: B

29. Most oxygen in the blood is transported
A) as deoxyhemoglobin.
B) as gas dissolved in plasma.
C) as oxyhemoglobin.
D) as carboxyhemoglobin.
E) as reduced hemoglobin.
Answer: C

30. Nerve impulses travel from the active inspiratory area to the diaphragm via the
A) vagus nerve.
B) phrenic nerve.
C) thoracic nerve.
D) intercostals nerve.
E) radial nerve.
Answer: B

31. Approximately how much CO_2 in the blood is carried as bicarbonate?
A) less than 40%
B) 50%
C) close to 60%
D) less than 70%
E) almost 80%
Answer: E

32. When stretch receptors in the lungs are activated
A) expiration will occur.
B) the lungs will deflate.
C) the inspiratory area is inhibited.
D) impulses are sent along the vagus nerve.
E) All of the above.
Answer: E

33. The primary chemical stimulus for breathing is the concentration of
A) carbon monoxide in the blood.
B) H^+ and carbon dioxide concentration of blood.
C) the carbon dioxide concentration of the blood.
D) the oxygen concentration of the blood.
E) the carbonic acid concentration in the blood.
Answer: B

34. The basic rhythm of respiration is controlled by the
A) pons.
B) medulla oblongata.
C) hypothalamus.
D) pneumotaxic area.
E) apneustic area.
Answer: B

35. Painful or labored breathing is referred to as
   A) rhinitis.
   B) diptheria.
   C) dyspnea.
   D) orthopena.
   E) epistaxis.
   Answer: C

36. A disorder characterized by the destruction of the alveolar walls is
   A) chronic bronchitis.
   B) emphysema.
   C) chronic obstructive pulmonary disease.
   D) asthma.
   E) tuberculosis.
   Answer: B

37. Respiration rates are controlled mainly by the
   A) medulla oblongata & midbrain and the amount of O_2 in the blood.
   B) medulla oblongata & pons and the amount of CO_2 in the blood.
   C) medulla oblongata & midbrain and the amount of CO_2 in the blood.
   D) medulla oblongata & pons and the amount of O_2 in the blood.
   E) medulla oblongata & hypothalamus and the amount of O_2 in the blood.
   Answer: B

38. The vocal cords are located in the
   A) esophagus.
   B) trachea.
   C) larynx.
   D) nasopharynx.
   E) laryngopharynx.
   Answer: C

39. Cartilaginous rings
   A) make the trachea rigid to keep the airway open.
   B) make the bronchi rigid to keep them open.
   C) are C-shaped in the trachea to allow the esophagus to expand behind it.
   D) are not found in the esophagus.
   E) All of the above are true.
   Answer: E

40. Breathing is controlled by the respiratory center(s) located in the
   A) cerebellum.
   B) medulla oblongata and pons.
   C) midbrain and hypothalamus.
   D) cerebrum.
   E) cerebellum and cerebrum.
   Answer: B

41. The flap of cartilage that closes off the larynx and prevents food and water from entering the larynx during swallowing is the
   A) epiglottis.
   B) glottis.
   C) thyroid cartilage.
   D) laryngocartilage.
   E) cricoid cartilage.
   Answer: A

42. Which of the following is not a function of the respiratory system?
    A) transport of nutrients to tissue
    B) ventilation of lungs
    C) Gas exchange between lungs and blood
    D) gas exchange between blood and cells
    E) production of sound
    Answer: A

43. Which of the following communicates with the laryngopharynx?
    A) larynx
    B) oropharynx
    C) esophagus.
    D) All of the above.
    E) None of the above.
    Answer: D

True-False

Write T if the statement is true and F if the statement is false.

1. The first step of respiration is external respiration.
   Answer: False

2. The nostrils are also called external nares.
   Answer: True

3. The middle portion of the pharynx is the nasopharynx.
   Answer: False

4. The mucous membrane of the larynx forms two pairs of folds.
   Answer: True

5. The trachea is located lateral to the esophagus.
   Answer: False

6. Tertiary bronchi divide into terminal bronchioles.
   Answer: False

7. The narrow top portion of the lung is called the apex.
   Answer: True

8. The right lung is divided into three lobes.
   Answer: True

9. In order for respiration to occur, the volume of the lung needs to be increased.
   Answer: True

10. The nasal cavity is divided into top, middle and bottom portions by nasal conchae and into a right and left side by the nasal septum. The purpose is to circulate air to warm, cleanse, examine and moisten it.
    Answer: True

11. The pressure inside the lungs is the alveolar pressure.
    Answer: True

12. The record of pulmonary volumes and capacities is called a spirogram.
    Answer: True

13. In clinical practice the word ventilation means inspiration only.
    Answer: False

14. The total pressure of a gas mixture is calculated by multiplying the partial pressures.
    Answer: False

15. The transport of respiratory gases between the lungs and body tissues is a function of the blood.
    Answer: True

16. During inspiration the diaphragm contracts and drops which increases the thoracic cavity's size and decreases its pressure. Expiration is the reverse, volume decreases and pressure increases forcing air out.
    Answer: True

17. If alveolar PCO_2 is low, CO_2 will diffuse from the capillary blood into the alveoli.
    Answer: True

18. The right lung is smaller than the left lung because of the position of the heart.
    Answer: False

19. Parietal pleural membrane lines the thoracic cavity while the visceral pleural membrane covers the lung.
    Answer: True

Short Answer

Write the word or phrase that best completes each statement or answers the question.

1. The lowest portion of the pharynx is the _____.
   Answer: laryngopharynx

2. The branches of the trachea to the bronchi and bronchioles is referred to as the _____.
   Answer: bronchial tree

3. The membrane that encloses and protects the lungs is the _____.
   Answer: pleural membrane

4. Terminal bronchioles subdivide into microscopic branches called _____.
   Answer: respiratory bronchioles

5. The maneuver used to expel an aspirated object is called the _____ maneuver.
   Answer: Heimlich

6. The respiratory gases move across the _____ membrane.
   Answer: alveolar-capillary (respiratory)

7. The visual examination of bronchi through a bronchoscope is called _______.
   Answer: bronchoscopy

8. When the diaphragm contracts it _____.
Answer: flattens

9. The term applied to normal quiet breathing is _____.
Answer: eupnea

10. The phospholipids produced by the alveolar cells are called _____.
Answer: surfactant

11. The total volume of air taken in during one minute is called the _____.
Answer: minute volume of respiration

13. The air that remains in the lungs after the expiratory reserve volume is expelled is the _____.
Answer: residual volume

14. The sum of residual volume plus expiratory reserve volume is the _____.
Answer: functional residual capacity

15. The _____ area controls the basic rhythm of respiration.
Answer: medullary rhythmicity

16. The passive process by which air flows into and out of the lungs is called ______.
Answer: ventilation (breathing)

17. The protective mechanism that prevents overinflation of the lungs is called _____.
Answer: inflation reflex

18. A slow rate and depth of respiration is called _____.
Answer: hypoventilation

19. The temporary cessation of breathing is known as _____.
Answer: apnea

20. The structure that prevents food from entering the respiratory passages is the _____.
Answer: epiglottis

21. The chemosensitive area is located in the _____.
Answer: medulla oblongata

22. Carbon dioxide can be carried by hemoglobin as _____.
Answer: carbaminohemoglobin

23. The immediate increase in ventilation at the onset of exercise is a result of the stimulation of _______.
Answer: proprioceptors

24. A chronic or acute inflammation of the mucous membranes in the nose is called ______.
Answer: rhinitis

Matching

Choose the item from Column 2 that matches each item in Column 1.

1. Column 1: The alveolar walls lose their elasticity and remain filled with air during expiration.
Column 2: emphysema

2. Column 1: Also called hyaline membrane disease
   Column 2: respiratory distress syndrome

3. Column 1: Kills infants between the ages of 1 week and 12 months, without warning
   Column 2: sudden infant death syndrome

4. Column 1: The presence of a blood clot in a pulmonary arterial vessel
   Column 2: pulmonary embolism

5. Column 1: An abnormal accumulation of interstitial fluid in the interstitial spaces and alveoli of the lungs
   Column 2: pulmonary edema

6. Column 1: Another term for nosebleed
   Column 2: epistaxis

7. Column 1: Spitting of blood from the respiratory tract
   Column 2: hemoptysis

8. Column 1: Oxygen starvation
   Column 2: asphyxia

9. Column 1: Painful breathing
   Column 2: dyspnea

10. Column 1: Difficult breathing that occurs in the horizontal position
    Column 2: orthopnea

Essay

Write your answer in the space provided or on a separate sheet of paper.

1. Name, in correct order, the conducting portion of the respiratory system.
   Answer: nose, pharynx, larynx, trachea, bronchi, bronchioles, terminal bronchioles

2. Name the function of the interior structures of the nose.
   Answer:
   1. They warm, moisten, and filter incoming air.
   2. They receive olfactory stimuli.
   3. They provide a resonating chamber for speech sounds.

3. Name and briefly describe the three basic processes of respiration.
   Answer:
   1. Pulmonary ventilation is the movement of air in and out of the lungs, inspiration occurs when the pressure inside the lungs is lower than the atmospheric pressure. This is achieved by increasing the volume of the lungs. Expiration occurs when the pressure in the lungs is greater than the pressure of the atmosphere. This occurs when the size of the thoracic cavity decreases.
   2. External respiration is the exchange of gases between the air in the alveoli and the blood.
   3. Internal respiration is the exchange of gases between the blood and tissues.

4. Define asthma.

   Answer: Asthma is a reaction, usually allergic, characterized by attacks of wheezing and difficult breathing, spasms of the smooth muscles in the walls of smaller bronchi, and bronchioles cause the passageways to close partially, resulting in an attack.

CHAPTER 19 The Digestive System

Multiple-Choice

Choose the one alternative that best completes the statement or answers the question.

1. All of the following are part of the gastrointestinal tract except the
   A) stomach.
   B) gallbladder.
   C) esophagus.
   D) small intestine.
   E) pharynx.
   Answer: B

2. Which of the following is NOT an accessory structure of the digestive system?
   A) liver
   B) gallbladder
   C) pancreas
   D) spleen
   E) teeth
   Answer: D

3. The layer of the GI tract wall that contains blood and lymphatic vessels is the
   A) muscularis mucosa.
   B) mucosa.
   C) submucosa.
   D) serosa.
   E) muscularis.
   Answer: C

4. The ability of the GI tract to mix and move material along its length is called
   A) digestion.
   B) ingestion.
   C) absorption.
   D) motility.
   E) excitability.
   Answer: D

5. The greater omentum is part of the
   A) mucosa.
   B) liver.
   C) mesentery.
   D) peritoneum.
   E) large intestine.
   Answer: D

6. Enteric neurons are found in the
   A) mucosa.
   B) submucosa.
   C) muscularis.
   D) serosa.
   E) Both A and B.
   Answer: B

7. The projection hanging from the soft palate is/are the
   A) papillae.
   B) uvula.
   C) lingual frenulum.
   D) fauces.
   E) tonsils.
   Answer: B

8. Which of the following contains taste buds?
   A) papillae
   B) uvula
   C) tonsils
   D) fauces
   E) none of the above
   Answer: A

9. Mumps is an inflammation of the
   A) sublingual glands.
   B) submandibular glands.
   C) parotid glands.
   D) thyroid glands.
   E) parathyroid glands.
   Answer: C

10. The tongue is composed of
   A) smooth muscle.
   B) skeletal muscle.
   C) loose connective tissue.
   D) glandular tissue.
   E) None of the above.
   Answer: B

11. Salivary amylase secreted into the oral cavity starts the digestion of
   A) proteins.
   B) carbohydrates.
   C) lipids.
   D) amino acids.
   E) glucose.
   Answer: B

12. Teeth are primarily composed of a bone-like substance called
   A) crown.
   B) enamel.
   C) cementum.
   D) gingivae.
   E) dentin.
   Answer: E

13. The dentin of the root is covered by
   A) crown.
   B) enamel.
   C) cementum.
   D) gingivae.
   E) dentin.
   Answer: C

14. The teeth which are adapted to cutting into food are the
    A) incisors.
    B) cuspids.
    C) premolars.
    D) molars.
    E) All of the above.
    Answer: A

15. All of the following are areas of the stomach EXCEPT
    A) cardia.
    B) duodenum.
    C) fundus.
    D) body.
    E) pylorus.
    Answer: B

16. Pepsinogen is produced by the
    A) hepatic cells in the liver.
    B) chief cells of the stomach.
    C) parietal cells of the stomach.
    D) exocrine glands of the pancreas.
    E) cells of the duodenum.
    Answer: B

17. The mucosa of the stomach, when empty, lies in large folds called
    A) gastric pits.
    B) gastric glands.
    C) rugae.
    D) peritoneum.
    E) omentum.
    Answer: C

18. The digestion of proteins by peptides starts in the
    A) oral cavity.
    B) esophagus.
    C) stomach.
    D) duodenum.
    E) ileum.
    Answer: C

19. The hormone that inhibits gastric emptying is
    A) gastric inhibitory protein.
    B) insulin.
    C) secretin.
    D) cholecystokinin.
    E) renin.
    Answer: D

20. Chyme is released by the stomach into the
    A) duodenum.
    B) esophagus.
    C) jejunum.
    D) ileum.
    E) cecum.
    Answer: A

21. Which of the following substances can be absorbed by the stomach?
   A) amino acids
   B) nucleic acids
   C) glucose
   D) fatty acids
   E) alcohol
   Answer: E

22. The G cells of the stomach secrete
   A) pepsinogen.
   B) pepsin.
   C) HCl.
   D) gastrin.
   E) CCK.
   Answer: D

23. The feeling of satiety is caused by
   A) cholecystokinin.
   B) secretin.
   C) pepsin.
   D) gastrin.
   E) All of the above.
   Answer: A

24. The pancreatic duct transports secretions from the pancreas to the
   A) stomach.
   B) liver.
   C) duodenum.
   D) jejunum.
   E) cecum.
   Answer: C

25. All of the following are substances found in pancreatic juice EXCEPT
   A) pepsin.
   B) trypsin.
   C) chymotrypsin.
   D) ribonuclease.
   E) lipase.
   Answer: A

26. Which of the following is NOT a protein-digesting enzyme?
   A) trypsin
   B) chymotrypsin
   C) carboxypeptidase
   D) amylase
   E) elastase
   Answer: D

27. Bile is produced by
   A) lymphocytes.
   B) Kupffer's cells.
   C) parietal cells.
   D) hepatocytes.
   E) chief cells.
   Answer: D

28. Substances absorbed by the small intestine are carried in the liver by the
    A) hepatic artery.
    B) hepatic vein.
    C) hepatic portal vein.
    D) common hepatic duct.
    E) central vein.
    Answer: C

29. All of the following are functions of the liver EXCEPT
    A) red blood cell production.
    B) storage of vitamins.
    C) synthesis of bile salts.
    D) excretion of bilirbin.
    E) activation of Vitamin D.
    Answer: A

30. Most digestion and absorption occur in the
    A) oral cavity.
    B) esophagus.
    C) stomach.
    D) small intestine.
    E) large intestine.
    Answer: D

31. The final portion of the small intestine is the
    A) ileum.
    B) duodenum.
    C) cecum.
    D) jejunum.
    E) colon.
    Answer: A

32. Which of the following is NOT an enzyme in the breakdown of carbohydrates?
    A) pancreatic amylase
    B) salivary amylase
    C) maltase
    D) peptidase
    E) lactase
    Answer: D

33. Absorption in the small intestine occurs by
    A) diffusion.
    B) osmosis.
    C) facilitated diffusion.
    D) active transport.
    E) All of the above.
    Answer: E

34. Which of the following are absorbed into lacteals?
    A) monosaccharides
    B) amino acids
    C) triglycerides
    D) nucleic acids
    E) all of the above
    Answer: C

35. The first portion of the large intestine is the
  A) colon.
  B) cecum.
  C) anal canal.
  D) rectum.
  E) ileum.
  Answer: B

36. The appendix is attached to the
  A) cecum.
  B) ileum.
  C) ascending colon.
  D) descending colon.
  E) rectum.
  Answer: A

37. The large intestine absorbs
  A) water.
  B) sodium.
  C) chloride.
  D) vitamins.
  E) All of the above.
  Answer: E

38. The portion of the large intestine just before the rectum is the
  A) anus.
  B) cecum.
  C) ascending colon.
  D) transverse colon.
  E) sigmoid colon.
  Answer: E

39. The exocrine protein of the pancreas
  A) is made of acinar cells.
  B) secretes several different enzymes to digest the energy nutrients.
  C) is controlled by the parasympathetic nervous system.
  D) connects to the alimentary canal by the pancreatic duct.
  E) All of the above are true about the exocrine portion of the pancreas.
  Answer: E

40. Cholecystokinin (CCK)
  A) stimulates the gall bladder to contract.
  B) is released in response to lipids entering the duodenum.
  C) stimulates the release of pancreatic juice.
  D) slows the movement of chyme into the duodenum.
  E) All of the above are true of CCK.
  Answer: E

41. Which of the following does not release substances into the duodenum?
  A) stomach
  B) spleen
  C) gall bladder
  D) pancreas
  E) liver
  Answer: B

42. Most of the nutrients have been absorbed from the chime by the time it reaches the
   A) cardiac region of stomach.
   B) duodenum.
   C) jejunum.
   D) ascending colon.
   E) pyloric region of stomach.
   Answer: D

43. Which of the following is the correct order of movement of substances through the large intestine?
   A) ilieum, cecum, descending colon, transverse colon, ascending colon, sigmoid colon, rectum
   B) cecum, ascending colon, transverse colon, descending colon, sigmoid colon, rectum
   C) cecum, descending colon, transverse colon, ascending colon, sigmoid colon, rectum
   D) sigmoid colon, ascending colon, transverse colon, descending colon, cecum, rectum
   E) transverse colon, ascending colon, sigmoid colon, descending colon, cecum, rectum
   Answer: B

44. The organ that stores fat, glycogen, iron and certain minerals, and detoxifies certain chemicals is the
   A) colon.
   B) pancreas.
   C) liver.
   D) stomach.
   E) spleen.
   Answer: C

45. The __ is a 20 foot piece of peritoneum that keeps the small intestines from springing out all over.
   A) ileocecal valve
   B) mesentery
   C) lesser omentum
   D) pleura
   E) pyloric sphinter
   Answer: B

True-False

Write T if the statement is true and F if the statement is false.

1. The inner lining of the GI tract is a mucous membrane.
   Answer: True

2. The greater omentum contains many lymph nodes.
   Answer: True

3. The tongue is part of the GI tract.
   Answer: False

4. The lingual tonsil lies at the base of the tongue.
   Answer: True

5. The submandibular glands lie in front of the sublingual glands.
   Answer: False

6. Sympathetic stimulation results in dryness of the mouth.
   Answer: True

7. The enzyme salivary amylase starts the breakdown of proteins in the mouth.
   Answer: False

8. The esophagus lies behind the trachea.
   Answer: True

9. The muscularis of the stomach has three layers.
   Answer: True

10. Parasympathetic stimulation inhibits digestion.
    Answer: False

11. The presence of chyme in the duodenum slows gastric emptying.
    Answer: True

12. The liver is the heaviest gland of the body.
    Answer: True

13. The gallbladder produces bile.
    Answer: False

14. Chief cells are found in the large intestine.
    Answer: False

15. Most of the energy nutrients are small enough to be absorbed in the small intestine by the blood capillaries but the nutrient, fatty acids, must be absorbed into lacteals.
    Answer: True.

16. Both food and air pass through the oropharynx so the uvula protects the nasal cavity when swallowing.
    Answer: True.

17. The mucosal layer of the large intestine is lined with millions of fingerlike projections called villi.
    Answer: False.

18. Many things can stimulate the vomit reflex including irritation, overextension of the stomach and various chemicals.
    Answer: True.

19. The cardiac sphincter prevents food and acid from moving into the duodenum too quickly.
    Answer: False

Short Answer

Write the word or phrase that best completes each statement or answers the question.

1. The medical specialty that deals with the structure, function, diagnosis, and treatment of the stomach and intestines is called ____.
   Answer: gastroenterology

2. The physiological term for eating is _______.
   Answer: ingestion

3. The waves of muscular contraction in the GI tract are referred to as ____.
   Answer: peristalsis

4. The mouth opens into the oropharynx through an opening called the ____.
Answer: fauces

5. The food in the mouth is shaped into a rounded mass called the ____.
Answer: bolus

6. The dentin of the crown of a tooth is covered by ____.
Answer: enamel

7. Swallowing is divided into tree stages: the voluntary stage, the pharyngeal stage, and the ____ stage.
Answer: esophageal

8. The large, central portion of the stomach is called the ____.
Answer: body

9. Hydrochloric acid of the stomach is produced by ____.
Answer: parietal cells

10. The enteric nervous system is found in the _______.
Answer: submucosa

11. The intestinal hormone, which decreases gastric secretion, is ____.
Answer: secretin

12. The acini of the pancreas produce ____.
Answer: pancreatic juice

13. The common hepatic duct joins the cystic duct to form the ____.
Answer: common bile duct

14. The principal bile pigment is ____.
Answer: bilirubin

15. The hormone responsible for the ejection of bile from the gallbladder is ____.
Answer: cholecystokinin (CCK)

16. The mucosa of the small intestine forms finger-like structures called ____.
Answer: villi

17. Each villus of the small intestine has a lymphatic vessel called a ____.
Answer: lacteal

18. The movements of the small intestine are peristalsis and ____.
Answer: segmentation

19. Protein digestion starts with pepsin and is completed by the enzyme ____.
Answer: peptidases

20. Bile salts act as emulsifying agents forming tiny spheres called ____.
Answer: micelles

21. The inactive form of trypsin is activated by an enzyme called ______.
Answer: enterokinase

22. The puckered appearance of the colon is caused by a series of pouches called ____.
    Answer: haustra

23. The last stage of digestion is carried out by ____.
    Answer: bacteria

24. Kupffer's cells are found in the _____.
    Answer: liver

25. An inflammation of the liver is called ______.
    Answer: hepatitis

Matching

Choose the item from Column 2 that best matches each item in Column 1.

1. Column 1: Inner lining of the GI tract
   Column 2: mucosa

2. Column 1: Outermost layer of the GI tract
   Column 2: serosa

3. Column 1: Secrete the hormone gastrin
   Column 2: G cells

4. Column 1: Produce hydrochloric acid
   Column 2: parietal cells

5. Column 1: Secrete pepsinogen
   Column 2: chief cells

6. Column 1: Secrete lysozyme
   Column 2: paneth cells

7. Column 1: "Good" cholesterol
   Column 2: high-density lipoproteins

8. Column 1: "Bad" cholesterol
   Column 2: low-density lipoproteins

9. Column 1: A type of food poisoning caused by a bacterial toxin.
   Column 2: Botulism

10. Column 1: A viral disease
    Column 2: Hepatitis

Essay

Write your answer in the space provided or on a separate sheet of paper.

1. Name and describe the five basic activities of the digestive system.
   Answer: 1) Ingestion- The take-in of food through the mouth
   2) Mixing and movement of food along the digestive tract through muscle movement.
   3) Digestion- The chemical and mechanical breakdown of food.

4) Absorption- The uptake of digested food from the GI tract into the circulation.
5) Defecation- The elimination of indigestible substances from the GI tract.

2. Name and briefly describe the location of the different salivary glands.
   Answer: 1) Parotid glands are located under and in front of the ears between the skin and the masseter muscle.
   2) The submandibular glands lie beneath the base of the tongue in the floor of the mouth.
   3) The sublingual glands lie in front of the submandibular glands.

3. Name the functions of the liver.
   Answer: 1. Carbohydrate metabolism
   2. Lipid metabolism
   3. Protein metabolism
   4. Removal of drugs and hormones
   5. Excretion of bile
   6. Synthesis of bile salts
   7. Storage
   8. Phagocytosis
   9. Activation of Vitamin D

4. Briefly describe the absorption of nutrients in the small intestine.
   Answer: Monosaccharides, amino acids, and short-chain fatty acids are absorbed into blood capillaries. Long-chain fatty acids and monoglycerides are absorbed as part of micelles, re-synthesized to triglycerides, and transported as chylomicrons. Chylomicrons are taken up by the lacteals.

CHAPTER 20 Nutrition and Metabolism

Multiple-Choice

Choose the one alternative that best completes the statement or answers the question.

1. Minerals constitute about __ % of the total body weight.
   A) 2
   B) 4
   C) 6
   D) 8
   E) 10
   Answer: B

2. Each gram of dietary protein provides about 4 calories, a gram of fat provides about ___ calories.
   A) 5
   B) 7
   C) 9
   D) 11
   E) 13
   Answer: C

3. The vitamin essential to the absorption and utilization of calcium and phosphorus is Vitamin ____.
   A) C
   B) E
   C) A
   D) B
   E) D
   Answer: E

4. The vitamin essential for the formation of photopigments of the retina is Vitamin ____.
   A) C
   B) E
   C) A
   D) B
   E) D
   Answer: C

5. The term metabolism refers to
   A) anabolic reactions.
   B) catabolic reactions.
   C) oxidation.
   D) reduction.
   E) All chemical reactions of the body.
   Answer: E

6. The breakdown of complex organic molecules into smaller ones is known as
   A) anabolism.
   B) catabolism.
   C) metabolism.
   D) oxidation.
   E) reduction.
   Answer: B

7. A chemical reaction that requires energy is
   A) anabolism.
   B) catabolism.
   C) metabolism.
   D) oxidation.
   E) All of the above.
   Answer: A

8. Chemically, enzymes are
   A) carbohydrates.
   B) proteins.
   C) lipids.
   D) nucleic acids.
   E) simple sugars.
   Answer: B

9. Each gram of a carbohydrate produces about
   A) 2.0 kilocalories.
   B) 4.0 kilocalories.
   C) 6.0 kilocalories.
   D) 8.0 kilocalories.
   E) 10. 0 kilocalories.
   Answer: B

10. Glucose is stored in the liver as
    A) starch.
    B) cellulose.
    C) fat.
    D) glycogen.
    E) ATP.
    Answer: D

11. The rate of glucose transport across cell membranes is greatly increased by
    A) insulin.
    B) glucagon.
    C) thyroxin.
    D) ADH.
    E) calcium.
    Answer: A

12. The net gain of ATP for each molecule of glucose which undergoes glycolysis is
    A) 10.
    B) 8.
    C) 6.
    D) 4.
    E) 2.
    Answer: E

13. Where in a cell does the Krebs cycle occur?
    A) nucleus
    B) cytoplasm
    C) mitochondria
    D) ribosomes
    E) Golgi apparatus
    Answer: C

14. Before entering the Krebs cycle, pyruvic acid has to be changed and combined with
    A) ATP.
    B) CoA.
    C) NADH.
    D) FADH.
    E) GTP.
    Answer: B

15. The largest amount of ATP molecules in cellular respiration are produced during the
    A) Krebs cycle.
    B) electron transport chain.
    C) glycolysis.
    D) transition reaction.
    E) fermentation.
    Answer: B

16. Glycogenolysis occurs when
    A) blood levels of glucose drop.
    B) glucagon is released.
    C) blood levels of glucose are high.
    D) Both A and B.
    E) Both B and C.
    Answer: D

17. The formation of glucose from proteins and fats is called
    A) glycogenolysis.
    B) lipolysis.
    C) glycogenesis.
    D) gluconeogenesis.
    E) lipogenesis.
    Answer: D

18. Ketone bodies are the result of catabolism of
    A) glycerol.
    B) fatty acids.
    C) amino acids.
    D) glycogen.
    E) glucose.
    Answer: B

19. Before entering the Krebs cycle,, fatty acids are converted into molecules of
    A) glucose.
    B) glyceraldehyde-3-phosphate.
    C) pyruvic acid.
    D) acetyl CoA.
    E) fructose-6-phosphate.
    Answer: D

20. Excess carbohydrates are synthesized into
    A) amino acids.
    B) proteins.
    C) triglycerides.
    D) ketone bodies.
    E) None of the above.
    Answer: C

21. During digestion, proteins are broken down into molecules of
    A) glucose.
    B) fatty acids.
    C) glycerol.
    D) amino acids.
    E) nucleic acids.
    Answer: D

22. Protein synthesis occurs in
    A) ribosomes.
    B) mitochondria.
    C) lysosomes.
    D) Golgi complexes.
    E) smooth endoplasmic reticulum.
    Answer: A

23. Which of the following vitamins can be made by the body?
    A) Vitamin C & B
    B) Vitamin D & K
    C) Vitamin E & B
    D) Vitamin K & C
    E) Vitamin B's
    Answer: B

24. Which of the following vitamins is water-soluble?
    A) Vitamin B
    B) Vitamin A
    C) Vitamin D
    D) Vitamin E
    E) Vitamin K
    Answer: A

25. Which of the following factors affect the metabolic rate?
    A) exercise
    B) age
    C) gender
    D) hormones
    E) all of the above
    Answer: E

26. Which of the following hormones regulates the basal state of metabolism?
    A) vasopressin
    B) thyroxin
    C) insulin
    D) estrogen
    E) progesterone
    Answer: B

27. The transfer of heat from a warmer object to a cooler one without physical contact is called
    A) conduction.
    B) evaporation.
    C) convection.
    D) radiation.
    E) transformation.
    Answer: D

28. The thermostat of the human body is located in the
   A) thalamus.
   B) hypothalamus.
   C) cortex.
   D) medulla oblongata.
   E) limbic system.
   Answer: B

29. All of the following result in an increase of body temperature EXCEPT
   A) vasoconstriction.
   B) vasodilation.
   C) epinephrine release.
   D) release of thyroid hormones.
   E) shivering.
   Answer: B

30. In degrees of Celsius normal body temperature is near
   A) 35.
   B) 37.
   C) 39.
   D) 40.
   E) 41.
   Answer: B

31. Sympathetic stimulation
   A) increases cellular metabolism.
   B) decreases cellular metabolism.
   C) has no effect on cellular metabolism.
   D) decreases body temperature.
   E) cause vasodilation.
   Answer: A

32. Aspirin reduces fever by
   A) inhibiting the hypothalamus.
   B) inhibiting prostaglandin secretion.
   C) stimulating epinephrine release.
   D) decreasing metabolism.
   E) lowering thyroid function.
   Answer: B

33. Which of the following processes are a part of the carbohydrate metabolisms?
   A) glycolysis
   B) electron transport chain
   C) the Krebs cycle
   D) formation of acetyl coenzyme A
   E) All of the above
   Answer: E

34. Coenzyme A is a derivative of
   A) Vitamin D.
   B) Vitamin E.
   C) Vitamin B.
   D) Vitamin C.
   E) None of the above.
   Answer: C

35. How many molecules of ATP are generated during the complete oxidation of a glucose molecule?
    A) 30 to 32
    B) 32 to 34
    C) 34 to 36
    D) 36 to 38
    E) 38 to 40
    Answer: D

36. Which of the following compounds transport dietary lipids to adipose tissue for storage?
    A) chylomicrons
    B) low density lipoproteins
    C) high density lipoproteins
    D) peripheral proteins
    E) phospholipids
    Answer: A

37. Fatty acid metabolism begins in the
    A) cytoplasm.
    B) mitochondria.
    C) nucleus.
    D) smooth ER.
    E) ribosomes.
    Answer: B

38. Which of the following hormone(s) is/are the MAIN regulators of the basal metabolic rate?
    A) human growth hormone
    B) testosterone
    C) thyroid hormones
    D) insulin
    E) vasopressin
    Answer: C

39. Basal metabolic rate
    A) reflects the rate nutrients are metabolized when one is exercising.
    B) reflects nitrogen consumption.
    C) remains constant throughout the day.
    D) is the body's entire kcalorie requirement.
    E) None of the above are true of basal metabolism.
    Answer: E

40. The process of deamination refers to
    A) someone with an abnormally low basal metabolic rate.
    B) the removal of a nitrogen from an amino acid.
    C) the anabolism of amino acids into proteins.
    D) the increase in basal metabolic rate that comes from an excess of proteins.
    E) the anabolism of lipids from excess amino acids.
    Answer: B

41. Energy derived from nutrients is measured in
    A) ergs.
    B) kcalories.
    C) adipose tissue composition.
    D) newtons.
    E) BTU's.
    Answer: B

42. Which of the following does NOT describe minerals?
    A) they are inorganic
    B) they are needed in small quantities
    C) levels are controlled by hormones
    D) many help regulate enzymatic reactions
    E) they can be synthesized by the body
    Answer: E

43. Which of the following are examples of poor nutrition?
    A) iodine deficiency
    B) obesity
    C) vitamin toxicity
    D) kwashiorkor
    E) All are examples of poor nutrition.
    Answer: E

True-False

Write T if the statement is true and F if the statement is false.

1. The milk, yogurt, and cheese group form the base of the food pyramid.
   Answer: False

2. In anabolism, simple substances combine into more complex molecules.
   Answer: True

3. Many coenzymes are derivatives of vitamins.
   Answer: True

4. Catabolism is usually an energy-releasing reaction.
   Answer: True

5. The Krebs cycle is also called anaerobic cellular respiration.
   Answer: False

6. Glycogenesis decreases blood sugar levels.
   Answer: True

7. Approximately 80 percent of stored triglycerides are deposited in subcutaneous tissue.
   Answer: False

8. Liver cells synthesize lipids from glucose.
   Answer: True

9. Essential amino acids can be synthesized by body cells.
   Answer: False

10. Large quantities of vitamins are necessary to maintain normal metabolism.
    Answer: False

11. A kilocalorie is equal to 1000 calories
    Answer: True

12. The ingestion of food can raise the metabolic rate.
    Answer: True

13. Both vitamin K & D can be made in the human body, D on the skin and K by bacteria in the large intestine.
    Answer: True

14. Carbohydrates are the only nutrient whose catabolism begins with the process of glycolysis.
    Answer: True.

15. The production of new glucose from either lipids or amino acids is called glygenolysis.
    Answer: False

16. The most important number in a cholesterol check is a low amount of HDL because that is the "bad" cholesterol.
    Answer: False

17. To a point, fever is beneficial in fighting microbes.
    Answer: True

18. Because it is derived from amino acids, Nutrasweet is harmless to everyone.
    Answer: False

Short Answer

Write the word of phrase that best completes each statement or answers the question.

1. Organic nutrients that are required in small amounts to maintain normal metabolism are called _____.
   Answer: vitamins

2. Vitamins that can inactivate oxygen free radicals are called ______.
   Answer: antioxidants

3. The principal anion in extracellular fluid is ____.
   Answer: Cl^-

4. The vitamin produced by intestinal bacteria and essential in the production of prothrombin is Vitamin __.
   Answer: K

5. The formation of peptide bonds between amino acids is a(n) ____ process.
   Answer: anabolic

6. The removal of electrons in the conversion of lactic acid into pyruvic acid is an example of ____.
   Answer: oxidation

7. The oxidation of glucose into two molecules of pyruvic acid is called ____.
   Answer: glycolysis

8. The food guide pyramid suggests ___ to ___ servings of the vegetable group per day.
   Answer: 3 to 5

9. The series of oxidation-reduction reactions that take place on the inner membrane of the mitochondria, and generate a large number of ATP molecules, is known as ____.
   Answer: electron transport chain

10. The formation of glycogen from glucose is called ____.
    Answer: glycogenesis

11. The splitting of triglyceride molecules is a process called ____.
    Answer: lipolysis

12. Before fatty acids can be metabolized by cells they have to undergo a series of reactions in the liver called ____.
    Answer: beta-oxidation

13. The removal of nitrogen from amino acids is called ____.
    Answer: deamination

14. Amino acids that can be synthesized by body cells are referred to as ____.
    Answer: nonessential amino acids

15. Vitamins that are stored in hepatocytes are ____ -soluble.
    Answer: fat

16. The higher the body temperature, the ____ the metabolic rate.
    Answer: higher

17. The metabolic rate ____ with age.
    Answer: decreases

18. The transfer of heat by movement of a liquid or gas between areas of different temperature is called ____.
    Answer: convection

19. The neurons of the hypothalamus involved in the regulation of body temperature are located in the ____.
    Answer: preoptic area

20. The transfer of body heat to a substance or object in contact with the body is called ____.
    Answer: conduction

21. Thyroid hormones will ____ body temperature.
    Answer: increase

22. The phase of fever when the skin becomes warm and the person begins to sweat is called the ____.
    Answer: crisis

23. The abbreviation BMR means ____.
    Answer: basal metabolic rate

24. The condition when body weight is 10 to 20 percent above the desirable standard due to fat accumulation is called ____.
    Answer: obesity

25. ____ refers to chemical reactions in the body.
    Answer: metabolism

Matching

Choose the item from Column 2 that best matches each item in Column 1.

1. Column 1: Chemical reactions that require energy.
   Column 2: anabolic reactions

2. Column 1: Chemical reactions that release energy.
   Column 2: catabolic reactions

3. Column 1: The removal of electrons from a molecule.
   Column 2: oxidation

4. Column 1: The addition of electrons to a molecule.
   Column 2: reduction

5. Column 1: The formation of glucose from non-carbohydrates.
   Column 2: gluconeogenesis

6. Column 1: The conversion of glycogen to glucose.
   Column 2: glycogenolysis

7. Column 1: The formation of glycogen from glucose.
   Column 2: glycogenesis

8. Column 1: A water-soluble vitamin.
   Column 2: Vitamin C

9. Column 1: A fat-soluble vitamin.
   Column 2: Vitamin A

Essay

Write your answer in the space provided or on a separate sheet of paper.

1. Define nutrients and name the principal classes of nutrients.
   Answer: Nutrients are chemical substances in food that provide energy or assist in the functioning of various body processes. The principal classes of nutrients are carbohydrates, lipids, proteins, minerals, vitamins, and water.

2. Name the guidelines for healthy eating.
   Answer: 1) Eat a variety of foods
   2) Maintain healthy weight
   3) Choose foods low in fat, saturated fat, and cholesterol
   4) Eat plenty of vegetables, fruits, and grain products
   5) Use sugars only in moderation
   6) Use salt and sodium only in moderation
   7) If you drink alcoholic beverages, do so in moderation

3. Name the steps of cellular respiration
   Answer: glycolysis, the Krebs cycle, and the electron transport chain

4. Name the principal routes by which the body can loose heat.
   Answer: radiation, evaporation, convection, and conduction

CHAPTER 21 The Urinary System

Multiple-Choice

Choose the one alternative that best completes the statement or answers the question.

1. The kidneys
   A) help regulate blood volume.
   B) help to control blood pressure.
   C) secrete erythropoietin.
   D) help control blood pH.
   E) All of the above are correct.
   Answer: E

2. All of the following belong to the urinary system EXCEPT
   A) urethra.
   B) ureter.
   C) bladder.
   D) prostate.
   E) kidneys.
   Answer: D

3. Together with the skin and the liver, the kidney synthesize _____, the active form of Vitamin D.
   A) calcitonin
   B) calcitriol
   C) erythropoietin
   D) renin
   E) angiotensin
   Answer: B

4. The renal pyramids are located in the
   A) renal cortex.
   B) renal medulla.
   C) renal capsule.
   D) pelvis.
   E) papillae.
   Answer: B

5. The functional units of the kidneys are called
   A) glomeruli.
   B) calyces.
   C) nephrons.
   D) corpuscles.
   E) tubules.
   Answer: C

6. The blood vessel leading away from the glomerulus is the
   A) renal vein.
   B) renal artery.
   C) peritubular capillary.
   D) efferent arteriole.
   E) afferent arteriole.
   Answer: D

7. The Bowman's capsule is part of the
   A) renal pyramid.
   B) renal corpuscle.
   C) renal tubule.
   D) renal pelvis.
   E) minor calyx.
   Answer: B

8. The first portion of the renal tubule is the
   A) collecting duct.
   B) papillary duct.
   C) proximal convoluted tubule.
   D) distal convoluted tubule.
   E) loop of Henle.
   Answer: C

9. The distal convoluted tubules empty into a
   A) minor calyx.
   B) loop of Henle.
   C) papillary duct.
   D) pelvis.
   E) collecting tubule.
   Answer: E

10. The interlobar arteries divide into the
   A) interlobular arteries.
   B) afferent arterioles.
   C) segmental arteries.
   D) arcuate arteries.
   E) peritubular capillaries.
   Answer: D

11. The peritubular capillaries drain into the
   A) arcuate veins.
   B) interlobular veins.
   C) interlobar veins.
   D) segmental veins.
   E) renal vein.
   Answer: B

12. Which of the following represents the correct order of anatomical structures in the nephron?
   A) glomerulus, proximal convoluted tubules, loop of Henle, distal convoluted tubules
   B) proximal convoluted tubules, loop of Henle, distal convoluted tubules, glomerulus
   C) glomerulus, distal convoluted tubules, loop of Henle, proximal convoluted tubules
   D) distal convoluted tubules, glomerulus, proximal convoluted tubules, loop of Henle
   E) glomerulus, loop of Henle, proximal convoluted tubules, distal convoluted tubules
   Answer: A

13. The cells of the inner wall of the glomerulus are called
   A) macula densa.
   B) podocytes.
   C) glomerular cells.
   D) capsular cells.
   E) epithelial cells.
   Answer: B

14. On average ___ to ___ liters of urine are excreted per day.
    A) 1-2
    B) 2-3
    C) 3-4
    D) 4-5
    E) Less than all of the above.
    Answer: A

15. Most glucose molecules are reabsorbed in the
    A) proximal convoluted tubules.
    B) distal convoluted tubules.
    C) ascending limb of the loop of Henle.
    D) descending limb of the loop of Henle.
    E) collecting ducts.
    Answer: A

16. The first step in urine formation is
    A) tubular secretion.
    B) secretion of ADH.
    C) tubular reabsorption.
    D) water reabsorption.
    E) glomerular filtration.
    Answer: E

17. Which of the following is found in plasma but not usually in glomerular filtrate?
    A) nitrogenous wastes
    B) large proteins
    C) vitamins
    D) water
    E) amino acids
    Answer: B

18. The glomerular filtration rate in a normal adult male is about
    A) 250 liters a day.
    B) 210 liters a day.
    C) 180 liters a day.
    D) 120 liters a day.
    E) 80 liters a day.
    Answer: C

19. All of the following processes are involved in tubular reabsorption EXCEPT
    A) osmosis.
    B) diffusion.
    C) filtration.
    D) active transport.
    E) All of the above are part of tubular reabsorption.
    Answer: C

20. Which of the following substances can be eliminated from the blood by tubular secretions?
    A) potassium ions
    B) hydrogen ions
    C) ammonium ions
    D) penicillin
    E) All of the above.
    Answer: E

21. Which of the following diuretics inhibit ADH secretion?
A) coffee
B) tea
C) beer
D) wine
E) Both C and D.
Answer: E

22. From the collecting ducts urine drains into the
A) renal pelvis.
B) urinary bladder.
C) minor calyces.
D) major calyces.
E) ureter.
Answer: C

23. The structure that connects a kidney to the urinary bladder is the
A) ureter.
B) urethra.
C) renal pelvis.
D) collecting duct.
E) calyx.
Answer: A

24. Which of the following transport urine from the kidneys to the urinary bladder?
A) peristalsis
B) hydrostatic pressure
C) gravity
D) All of the above.
E) None of the above.
Answer: D

25. The urinary bladder is lined by
A) pseudostratified epithelium.
B) transitional epithelium.
C) sqamous epithelium.
D) cuboidal epithelium.
E) columnar epithelium.
Answer: B

26. Urine is expelled from the urinary bladder by the process of
A) excretion.
B) defecation.
C) micturition.
D) filtration.
E) None of the above.
Answer: C

27. The terminal portion of the urinary system is the
A) urinary bladder.
B) calyx.
C) ureter.
D) urethra.
E) collecting duct.
Answer: D

28. Which of the following can decrease urine flow?
   A) nervousness
   B) high temperatures
   C) diuretics
   D) increase water intake
   E) inhibition of ADH.
   Answer: B

29. All of the following are principle solutes of urine EXCEPT
   A) urea.
   B) creatinine.
   C) glycogen.
   D) uric acid.
   E) urobilinogen.
   Answer: C

30. Bedwetting is referred to as
   A) enuresis.
   B) anuria.
   C) diuresis.
   D) dysuria.
   E) azotemia.
   Answer: A

31. Which of the following hormones increase reabsorption of water in the collecting tubules?
   A) renin
   B) ADH
   C) aldosterone
   D) insulin
   E) angiotensin
   Answer: B

32. What is the percentage of water in normal urine?
   A) 95%
   B) 87%
   C) 80%
   D) 75%
   E) 50%
   Answer: A

33. The average capacity of the urinary bladder is ____ to ____ mL.
   A) 100-200
   B) 300-400
   C) 500-600
   D) 700-800
   E) 900-1000
   Answer: D

34. On average the renal blood flow declines by ___ between the ages of 40 and 70.
   A) 20%
   B) 30%
   C) 40%
   D) 50%
   E) 60%
   Answer: D

35. Which part of the kidney filters the fluids (blood)?
A) renal calyx
B) renal hilum
C) renal corpuscle
D) renal tubule
E) renal artery
Answer: C

36. Which is the correct path?
A) PCT to loop of Henle to collececting duct to calyx to ureter
B) DCT to loop of Henle to calyx to collecting duct to pelvis
C) DCT to PCT to calyx to ureter to pelvis
D) ACT to BCT to CCT to DCT
E) DCT to collecting duct to calyx to pelvis
Answer: E

37. Secretion of ADH promotes the
A) reabsorption of water.
B) excretion of water.
C) excretion of sodium molecules
D) secretion of sodium molecules.
E) All of the above.
Answer: A

38. Which of the following should not be found in urine?
A) urea
B) uric acid
C) sodium
D) glucose
E) H ions
Answer: D

38. Substances that do not pass from blood into the glomerular capsule include
A) positively charged ions like sodium.
B) negatively charged ions like chloride.
C) plasma proteins.
D) water molecules.
E) nitrogenous wastes.
Answer: C

39. The initial volume of filtrate into the glomerular capsules per day is about
A) 5 pints.
B) 5 quarts.
C) 5 liters.
D) 50 -90 liters.
E) between 130 and 180 liters.
Answer: E

40. Uric acid is formed from the metabolism of
A) creatine phosphate.
B) amino acids.
C) nucleic acids.
D) carbohydrates.
E) fatty acids.
Answer: C

True-False

Write T if the statement is true and F if the statement is false.

1. The kidneys help regulate the blood pressure.
   Answer: True

2. The renal columns are part of the renal pyramids.
   Answer: False

3. The renal corpuscle filters plasma into the renal tubule.
   Answer: True

4. The proximal convoluted tubules are continuous with the ascending limb of Henle.
   Answer: False

5. The major work of the urinary system is done by the nephron.
   Answer: True

6. The blood pressure in the kidneys is about 56mm Hg.
   Answer: False

7. The glomerular filtration rate is primarily regulated by the nervous system.
   Answer: False

8. Water retention increases blood volume and therefore blood pressure.
   Answer: True

9. The opening and closing of the internal urethral sphincter is involuntary.
   Answer: True

10. The urethra is the tube connecting a kidney with the urinary bladder.
    Answer: False

11. Urine volume is influenced by blood pressure.
    Answer: True

12. Constriction of the afferent arteriole decreases blood flow into the glomerulus, which increases GFR.
    Answer: False

13. The kidneys produce erythropoietin.
    Answer: True

14. Since the efferent arteriole is narrower than the afferent arteriol, fluids and other things are filtered through the glomerulus into the glomerular capsule.
    Answer: True

15. Like the digestive system, the urinary system has two sphincters at its end, the first involuntary and the second voluntary.
    Answer: True.

16. The blood travels in the kidney by this route: It enters the kidney through the renal artery, then several arteries into the afferent arteriole. From there the glomerulus filters it. The remaining blood leaves through the efferent arteriole to the peritubular capillarier to several venules and finally into the renal vein.
    Answer: True.

17. The kidney attempts to raise blood pressure because blood pressure is related to GFR. If filtration is occurring too rapidly, the kidney secretes hormones to lower blood pressure.
Answer: True

18. Urine/filtrate travels from the glomerulus to glomerular capsule to the proximal convoluted tubule and into the loop of Henle. Then it travels through the distal convoluted tubule to the collecting tubules which empty into the calyces and finally the renal pelvis. From there, urine goes through the ureters to urinary bladder and out the urethra.
Answer: True

Short Answer

Write the word or phrase that best completes each statement or answers the question.

1. A toxic level of urea in the blood is called ____.
Answer: uremia

2. Each kidney is enclosed in a transparent, fibrous membrane called _____.
Answer: renal capsule

3. The large cavity close to the dent in the kidney is the ____.
Answer: renal pelvis

4. Urine formation involves three processes: 1) glomerular filtration, 2) tubular reabsorption, and 3) ____.
Answer: tubular secretion

5. The ability of the kidneys to maintain a constant blood pressure despite changes in the systemic blood pressure is due to ____.
Answer: renal autoregulation

6. The hormone that promotes excretion of water and excretion of sodium is ____.
Answer: atrial natriuretic peptide

7. The maximum amount of a substance that can be reabsorbed by the renal tubules is called ____.
Answer: tubular maximum (renal threshold)

8. The presence of glucose in the urine is referred to as ____.
Answer: glycosuria

9. The process of filtering blood by artificial means is called ____.
Answer: hemodialysis

10. The small triangular area at the base of the urinary bladder is called ____.
Answer: trigone

11. A lack of voluntary control over micturition is referred to as ____.
Answer: incontinence

12. An analysis of the volume, physical and chemical properties of urine is called ____.
Answer: urinalysis

13. A progressive and generally irreversible decline in the glomerular filtration rate is called ____.
Answer: chronic renal failure

14. The functional unit of the kidney is a (n) ____.
Answer: nephron

15. The hormone that increases water reabsorption is _____.
    Answer: ADH (antidiuretic hormone)

16. The tube that discharges urine from the body is called the ____.
    Answer: urethra

17. Excessive urine is referred to as _____.
    Answer: polyuria

18. The scientific study of the kidney is called ______.
    Answer: nephrology

19. The amount of filtrate produced by the kidneys per minute is referred to as ___ ___ ___.
    Answer: glomerular filtration rate

20. The glomerular filtrate is also called ___ ___.
    Answer: tubular fluid

21. A condition that is the result of inadequate release of ADH is ____ ____.
    Answer: diabetes insipidus

Matching

Choose the item from Column 2 that best matches each item in Column 1.

1. Column 1: An inflammation of the kidney that involves glomeruli.
   Column 2: glomerulonephritis

2. Column 1: An inflammation of the renal pelvis and its calyces.
   Column 2: pyelitis

3. Column 1: A condition that refers to protein in the urine.
   Column 2: nephrotic syndrome

4. Column 1: A decrease or cessation of glomerular filtration.
   Column 2: renal failure

5. Column 1: Presence of urea or other nitrogenous elements in the blood.
   Column 2: azotemia

6. Column 1: Hernia of the urinary bladder.
   Column 2: cystocele

7. Column 1: Increased excretion of urine.
   Column 2: diuresis

8. Column 1: Painful urination.
   Column 2: dysuria

9. Column 1: Toxic levels of urea in the blood.
   Column 2: uremia

10. Column 1: Narrowing of the lumen of the ureter or urethra.
    Column 2: stricture

Essay

Write your answer in the space provided or on a separate sheet of paper.

1. Describe the functions of the kidneys.
   Answer: 1. They regulate blood volume and composition of the blood; they remove waste and excrete selected amounts of wastes to include hydrogen ions, which helps control the pH of blood.
   2. They help regulate blood pressure by secreting renin, an enzyme which activates the renin-angiotensin system.
   3. They contribute to metabolism by 1) performing gluconeogenesis during starvation, 2) secreting erythropoietin which stimulates red blood cell production, and 3) participating in calcitriol synthesis, the active form of Vitamin D.

2. Name the components of a nephron in correct order, starting with the glomerolus.
   Answer: glomerolus-Bowman's capsule-proximal convoluted tubules-descending limb of the Loop of Henle-ascending limb of the loop of Henle-distal convoluted tubules-collecting duct.

3. Name the actions of angiotensin II.
   Answer: 1. Vasoconstriction of arterioles
   2. Stimulation of aldosterone secretion by the adrenal cortex
   3. Stimulation of the thirst center in the hypothalamus
   4. Stimulation of ADH secretion from the posterior pituitary

4. Define glomerular filtration rate.
   Answer: The glomerular filtration rate (GFR) is the amount of filtrate formed in both kidneys every minute.

CHAPTER 22 Fluid, Electrolyte, and Acid-Base Balance

Multiple-Choice

Choose the one alternative that best completes the statement or answers the question.

1. The body fluid found within the cells is called
   A) plasma.
   B) extracellular fluid.
   C) interstitial fluid.
   D) intracellular fluid.
   E) water.
   Answer: D

2. In lean adults body fluid makes about ___-___% of total body mass.
   A) 30-40
   B) 45-50
   C) 55-60
   D) 65-70
   E) 85-90
   Answer: C

3. Blood plasma belongs to
   A) intracellular fluid.
   B) extracellular fluid.
   C) interstitial fluid.
   D) Both B and C.
   E) None of the above.
   Answer: B

4. Peritoneal fluids between serous membranes belongs to
   A) intracellular fluid.
   B) extracellular fluid.
   C) interstitial fluid.
   D) Both B and C.
   E) None of the above.
   Answer: B

5. Which of the following extracellular fluids is found in the ear?
   A) aqueous humor
   B) endolymph
   C) peritoneal fluid
   D) synovial fluid
   E) cerebrospinal fluid
   Answer: B

6. Synovial fluid is found in
   A) the inner ear.
   B) the middle ear.
   C) joints.
   D) serous membranes.
   E) Both A and B.
   Answer: C

7. Most solutes in body fluid are
   A) electrolytes.
   B) proteins.
   C) sugars.
   D) amino acids.
   E) lipids.
   Answer: A

8. The highest proportion of water in body weight is found in
   A) lean women.
   B) lean men.
   C) fat women.
   D) fat men.
   E) infants.
   Answer: E

9. Water is lost through the actions of the
   A) kidneys.
   B) gastrointestinal tract.
   C) lungs.
   D) skin.
   E) All of the above.
   Answer: E

10. The center for thirst is located in the
    A) medulla oblongata
    B) hypothalamus
    C) pituitary
    D) kidney
    E) GI tract
    Answer: B

11. Dehydration stimulates the center for thirst by
    A) decreased production of saliva.
    B) increased osmotic pressure.
    C) decreased blood volume.
    D) B and C are correct.
    E) A, B, and C are correct.
    Answer: E

12. Total fluid output of the body under normal conditions is approximately
    A) 200 ml/day.
    B) 2500 ml/day.
    C) 400 ml/day.
    D) 1500 ml/day.
    E) 4300 ml/day.
    Answer: B

13. All of the following are involved in regulation of fluid output EXCEPT
    A) ADH.
    B) ANP.
    C) Cholecystokinin.
    D) Renin.
    E) Aldosterone.
    Answer: C

14. Which of the following hormones is responsible for increased urinary excretion of Na^+?
   A) angiotensin II
   B) aldosterone
   C) antidiuretic hormone
   D) atrial natriuretic peptide
   E) epinephrine
   Answer: D

15. Release of ADH is stimulated by
   A) low osmotic pressure.
   B) alcohol consumption.
   C) increased blood volume.
   D) increased osmotic pressure of plasma.
   E) All of the above.
   Answer: D

16. Osmotic pressure is related to
   A) the number of particles in a solution.
   B) the size of particles in a solution.
   C) the weight of particles in a solution.
   D) All of the above.
   E) None of the above.
   Answer: A

17. The major difference between plasma and interstitial fluid is
   A) the osmolarity.
   B) the concentration of carbohydrates.
   C) the concentration of solutes.
   D) the concentration of protein anions.
   E) the concentration of water.
   Answer: D

18. The most abundant cation in extracellular fluid is
   A) calcium.
   B) sodium.
   C) potassium.
   D) magnesium.
   E) hydrogen.
   Answer: B

19. The level of sodium in the blood is controlled by
   A) aldosterone.
   B) ADH.
   C) ANP.
   D) A and B.
   E) All of the above.
   Answer: E

20. ANP is produced in the
   A) kidneys.
   B) adrenal glands.
   C) pituitary gland.
   D) hypothalamus.
   E) atria of the heart.
   Answer: E

21. All of the following are characteristics of hyponatremia EXCEPT
    A) intense thirst.
    B) muscular weakness.
    C) dizziness.
    D) hypotension.
    E) tachycardia.
    Answer: A

22. A lower than normal level of potassium is referred to as
    A) hyperkalemia.
    B) hypokalemia.
    C) hyponatremia.
    D) hypernatrimia.
    E) hypochloremia.
    Answer: B

23. The levels of calcium in the blood are regulated by
    A) ADH and parathyroid hormone.
    B) aldosterone and ADH.
    C) parathyroid hormone and calcitonin.
    D) calcitonin and renin.
    E) renin and aldosterone.
    Answer: C

24. Hypercalcemia is characterized by
    A) lethargy.
    B) bone pain.
    C) muscle cramps.
    D) numbness and tingling of fingers.
    E) Both A and B.
    Answer: E

25. Which of the following is the major extracellular anion?
    A) phosphate
    B) calcium
    C) magnesium
    D) chloride
    E) potassium
    Answer: D

26. Which of the following ions diffuses easily between extracellular and intracellular compartments?
    A) magnesium
    B) chlorine
    C) sodium
    D) phosphate
    E) potassium
    Answer: B

27. Under normal conditions, any fluid not reabsorbed by the venous end of the capillaries
    A) remains in the extracellular spaces.
    B) is absorbed by the surrounding cells.
    C) passes into lymphatic capillaries.
    D) is excreted in perspiration.
    E) diffuses to other capillary beds.
    Answer: C

28 The pH of the extracellular fluid such as blood plasma, of a healthy person is
A) acidic.
B) slightly acidic.
C) neutral.
D) slightly basic.
E) basic.
Answer: D

29. The acid-base balance depends on the concentration of
A) calcium ions.
B) hydrogen ions.
C) potassium ions.
D) sodium ions.
E) magnesium ions.
Answer: B

30. A substance that dissociates into hydroxyl ions is a(n)
A) buffer.
B) acid.
C) base.
D) salt.
E) ion.
Answer: C

31. Which of the following buffer systems help regulate the pH in blood?
A) carbonic acid buffer system
B) phosphate buffer system
C) hemoglobin buffer system
D) protein buffer system
E) bicarbonate buffer system
Answer: D

32. The most abundant buffer system in body cells and plasma is the
A) carbonic acid buffer system.
B) bicarbonate buffer system.
C) phosphate buffer system.
D) hemoglobin buffer system.
E) protein buffer system.
Answer: E

33. An increase in the carbon dioxide concentration in body fluids results in
A) lower pH.
B) more acid fluids.
C) higher pH.
D) A and B.
E) B and C.
Answer: D

34. Increased respiration
A) removes carbon dioxide from the blood.
B) reduces the hydrogen ion concentration in blood.
C) increases the pH of blood.
D) All of the above.
E) None of the above.
Answer: D

35. The principal physiological effect of acidosis is
   A) excitation of the central nervous system.
   B) depression of the central nervous system.
   C) excitation of cardiovascular center.
   D) overexcitabilty of peripheral nerves.
   E) inhibition of the neuromuscular junctions.
   Answer: B

36. Which of the following compartments contains the most (over 60%) of the body's fluids?
   A) blood (plasma)
   B) tissue space (interstitial)
   C) cells (intracellular)
   D) joints (synovial)
   E) lymphatic (lymph)
   Answer: C

37. The two ways the body obtains water are:
   A) through the skin and drinking.
   B) ingestion and metabolism.
   C) urine output and sweating.
   D) defecation and ingestion.
   E) eating and drinking.
   Answer: B

38. Which of the following is not a considered fluid of the body?
   A) intracellular fluid
   B) cerebrospinal fluid
   C) blood and lymph
   D) interstitial fluid
   E) All are considered to be body fluids
   Answer: E

39. Which compartment has "first dibs" or priority for fluids if water becomes scarce in the body?
   A) tissue space
   B) blood
   C) lymph
   D) cell
   E) ventricles of the brain
   Answer: D

40. The kidney regulates the blood and body pH by
   A) reabsorbing hydrogen ions.
   B) excreting hydrogen ions.
   C) reabsorbing sodium ions.
   D) A & C.
   E) A & B.
   Answer: E

41. Substances in the body which prevent rapid changes in pH are
   A) carbohydrates.
   B) organic acids.
   C) inorganic acids.
   D) buffers
   E) organic bases.
   Answer: D

True-False

Write T if the statement is true and F if the statement is false.

1. Dehydration decreases the production of saliva.
   Answer: True

2. Excessive fluid in the blood results in hypotension.
   Answer: False

3. Glucose is one of the major electrolytes of the body.
   Answer: False

4. The ion concentration of extracellular fluid differs considerably from the ion concentration of intracellular fluids.
   Answer: True

5. A decrease in the sodium concentration causes the release of ADH from the posterior pituitary.
   Answer: False

6. Calcium is involved in blood clotting.
   Answer: True

7. Aldosterone is the main hormone in the homeostasis of calcium.
   Answer: False

8. Phosphate is mainly an intracellular electrolyte.
   Answer: True

9. Both body water and sodium are lost during excessive sweating.
   Answer: True

10. Lost body fluids should be replaced by drinking pure water.
    Answer: False

11. The majority of hydrogen ions are the result of cellular metabolism.
    Answer: True

12. Sodium is a strong acid and can be neutralized by bicarbonate.
    Answer: False

13. Respiration plays a role in the maintenance of blood pH.
    Answer: True

14. Both hydrostatic (a pushing pressure) and osmotic (a pulling pressure) are important in water and electrolyte balance.
    Answer: True

15. Both electrolytes and water can be lost through the processes of perspiration, defecation and urination.
    Answer: True

16. Using hypoventilation or hyperventilation to help balance pH is called respiratory compensation.
    Answer: True

17. Because acidosis results in an overexcitation of the neurons of the central nervous system, it can cause convulsions and muscle spasms.
Answer: False.

18. Taking diuretics has no effect on fluid or electrolyte balance.
Answer: False

Short Answer

Write the word or phrase that best completes each statement or answers the question.

1. Body fluids outside of cells are called ____.
Answer: extracellular fluid

2. When water loss is greater than water gain the result is ____.
Answer: dehydration

3. The thirst center is located in the ____.
Answer: hypothalamus

4. The hormone that increases fluid loss in the urine is ____.
Answer: atrial natriuretic peptide (ANP)

5. Negatively charged ions are called ____.
Answer: anions

6. Compounds that can conduct electric currents are called ____.
Answer: electrolytes

7. In osmosis, water moves from an area of ____ osmotic pressure to an area with ____ osmotic pressure.
Answer: lower, higher

8. A higher-than-normal blood sodium level is called ____.
Answer: hypernatremia

9. The levels of blood calcium are regulated by the hormones ____ and ____.
Answer: calcitonin, parathyroid hormone

10. The difference between the forces that move fluid out of the plasma and the forces that push it into the plasma is the ____.
Answer: net filtration pressure

11. The principal cation inside the cell is ____.
Answer: potassium

12. When carbon dioxide combines with water it forms ____.
Answer: carbonic acid

13. A substance that ionizes to produce hydrogen ions is a(n) ____.
Answer: acid

14. Systems that prevent drastic changes in the pH of bodily fluids are called ____.
Answer: buffer systems (buffers)

15. The inspiratory area is located in the _____.
    Answer: medulla oblongata

16. When blood pH decreases below 7.35, _____ occurs.
    Answer: acidosis

17. The physiological response to an acid-base imbalance to normalize blood pH is called ____.
    Answer: compensation

18. Plasma makes up ____% of the ECF.
    Answer: 20

19. The hormone responsible for decreased water loss in urine is _____.
    Answer: ADH

20. Substances that dissociate in water and form ions are ______.
    Answer: electrolytes

21. Renal compensation occurs in persons that have an altered blood pH due to ______ ____.
    Answer: respiratory causes

22. Water that is generated during the cell's chemical activity is called _____ water.
    Answer: metabolic

Matching

Choose the item from Column 2 that best matches each item in Column 1.

1. Column 1: muscular weakness, dizziness, headache, hypotension, tachycardia, and shock
   Column 2: hyponatremia

2. Column 1: intense thirst, fatigue, restlessness, agitation, and coma
   Column 2: hypernatremia

3. Column 1: cramps, fatigue, flaccid paralysis, nausea, vomiting, mental confusion, increased urine output, shallow respiration, and changes in the electrocardiogram
   Column 2: hypokalemia

4. Column 1: irritability, anxiety, abdominal cramping, diarrhea, weakness, and paresthesia
   Column 2: hyperkalemia

5. Column 1: numbness and tingling of the fingers, hyperactive reflexes, muscle cramps, tetany, and convulsions.
   Column 2: hypocalcemia

6. Column 1: lethargy, weakness, anorexia, nausea, vomiting, polyuria, itching, bone pain, depression, confusion, paresthesia, stupor, and coma.
   Column 2: hypercalcemia

7. Column 1: muscle spasms, alkalosis, depressed respiration, and coma
   Column 2: hypochloremia

Essay

Write your answer in the space provided or on a separate sheet of paper.

1. Name the major ions involved in the fluid and electrolyte balance and briefly describe their major location.

   Answer:
   1. Sodium is the most abundant extracellular ion and represents about 90 percent of extracellular cations.
   2. Potassium is the most abundant cation in the intracellular fluid.
   3. Calcium is the most abundant ion in the body and is principally an extracellular extracellular electrolyte.
   4. Magnesium is primarily an intracellular electrolyte.
   5. Chloride is the major extracellular anion.
   6. Phosphate is principally an intracellular electrolyte.

2. Explain the ways by which dehydration stimulates thirst.

   Answer:
   1. Dehydration decreases the production of saliva, which causes dryness of the oral mucosa. A dry mouth and pharynx stimulates the osmoreceptors of the hypothalamus, resulting in the sensation of thirst.
   2. Dehydration increases the osmotic pressure of blood, which also stimulates the osmoreceptors of the hypothalamus, resulting in thirst.
   3. Dehydration decreases the blood volume, which causes a drop in blood pressure. This stimulates the renin-angiotensin II system. Angiotensin II stimulates the thirst center of the hypothalamus.

3. Explain water intoxication.

   Answer: If lost bodily fluids are replaced by pure water, body fluids become more dilute, which causes the concentration of sodium to fall below normal. A decrease in the sodium concentration causes the osmotic pressure of the interstitial fluid to fall. The result is osmosis of water into intracellular fluid. The cells become hypotonic and swell, a condition called water intoxication. Water intoxication causes disorientation, convulsions, coma, and possible death.

CHAPTER 23 The Reproductive Systems

Multiple-Choice

Choose the one alternative that best completes the statement or answers the question.

1. Sperm production occurs in the
   A) Sertoli cells.
   B) seminiferous tubules.
   C) interstitial cells of Leydig.
   D) epididymis.
   E) vas deferens.
   Answer: B

2. Which of the following cells is the most immature?
   A) sperm cell
   B) spermatid
   C) spermatogonia
   D) primary spermatocyte
   E) secondary spermatocyte
   Answer: C

3. In the male, the hormone inhibin is produced in the
   A) hypothalamus.
   B) Sertoli cells.
   C) pituitary.
   D) interstitial cells of Leydig.
   E) spermatogonia.
   Answer: B

4. The cell produced by fertilization is called a(n)
   A) gamete.
   B) secondary oocyte.
   C) embryo.
   D) zygote.
   E) fetus.
   Answer: D

5. All of the following are true for meiosis EXCEPT
   A) it occurs during spermatogenesis.
   B) it occurs during oogenesis.
   C) the daughter cells have only half the chromosomes of the parent cell.
   D) it is subdivided into meiosis I and meiosis II.
   E) it can be used for cell repair.
   Answer: E

6. During spermatogenesis, the cells formed by Meiosis II are called
   A) spermatids.
   B) sperm cells.
   C) primary spermatocytes.
   D) secondary spermatocytes.
   E) spermatogonia.
   Answer: A

7. The production of testosterone in the interstitial cells of Leydig is stimulated by
   A) inhibin.
   B) luteinizing hormone.
   C) follicle-stimulating hormone.
   D) relaxin.
   E) progesterone.
   Answer: B

8. Sperm maturation occurs in the
   A) seminiferous tubules.
   B) straight tubules.
   C) epididymis.
   D) rete testes.
   E) vas deferens.
   Answer: C

9. The terminal duct of the male reproductive system is the
   A) ejaculatory duct.
   B) penis.
   C) urethra.
   D) ductus deferens.
   E) ureter.
   Answer: C

10. 60 percent of the volume of semen is produced by the
   A) prostate gland.
   B) Cowper's gland.
   C) seminal vesicles.
   D) testis.
   E) bulbourethral gland.
   Answer: C

11. The gland that surrounds the superior portion of the urethra in males is the
   A) Cowper's gland.
   B) prostate gland.
   C) bulbourethral gland.
   D) adrenal gland.
   E) None of the above.
   Answer: B

12. Semen production involves the
   A) testis.
   B) prostate gland.
   C) bulbourethral gland.
   D) seminal vesicles.
   E) All of the above.
   Answer: E

13. The structure of the sperm cell containing mitochondria is the
   A) acrosome.
   B) head.
   C) tail.
   D) midpiece.
   E) center.
   Answer: D

14. All of the following are structures of the penis EXCEPT
    A) corpus spongiosum.
    B) prepuce.
    C) external urethral orifice.
    D) glands penis.
    E) ejaculatory duct.
    Answer: E

15. The average volume of semen for an ejaculation is
    A) 1 to 2 mL.
    B) 1.5 to 3mL.
    C) 2.5 to 5mL.
    D) 4.5 to 6mL.
    E) 5 to 7 mL.
    Answer: C

16. Infertility in males becomes likely if the sperm count falls below
    A) 50 million sperm/mL.
    B) 40 million sperm/mL.
    C) 30 million sperm/mL.
    D) 20 million sperm/mL.
    E) 10 million sperm/mL.
    Answer: D

17. The enzyme of sperm cells necessary to penetrate an ovum is located in the
    A) head of the sperm cell.
    B) acrosome.
    C) midpiece.
    D) tail portion of the sperm cell.
    E) nucleus of the sperm.
    Answer: B

18. All of the following are true for testosterone EXCEPT
    A) it is an anabolic steroid.
    B) it stimulates the sex drive.
    C) it stimulates inhibin production.
    D) it is responsible for the secondary sex characteristics of males.
    E) All of the above are true.
    Answer: C

19. Spermatids contain _____ chromosomes.
    A) 16
    B) 20
    C) 23
    D) 36
    E) 46
    Answer: C

20. Sperm cells can survive up to ___ hours in the female reproductive tract.
    A) 10
    B) 24
    C) 36
    D) 48
    E) 52
    Answer: D

21. The ovarian follicle that ruptures during ovulation is the
    A) primary follicle.
    B) secondary follicle.
    C) tertiary follicle.
    D) Graafian follicle.
    E) None of the above.
    Answer: D

22. Fertilization usually occurs in the
    A) vagina.
    B) cervix.
    C) uterus.
    D) uterine tubes.
    E) ovaries.
    Answer: D

23. The layer of the uterine wall that is shed during menstruation is the
    A) endometrium.
    B) Cervical lining.
    C) perimetrium.
    D) myometrium.
    E) Both A and B.
    Answer: A

24. The external genitalia of the female are collectively called
    A) labia.
    B) vulva.
    C) clitoris.
    D) mons pubis.
    E) glans.
    Answer: B

25. The structure of the female reproductive system that contains erectile tissue is the
    A) eternal urethral orifice.
    B) vaginal orifice.
    C) labia minora.
    D) mons pubis.
    E) clitoris.
    Answer: E

26. All of the following are structures of the mammary glands EXCEPT
    A) Skene's gland.
    B) alveoli.
    C) Cooper's ligaments.
    D) areola.
    E) lobules.
    Answer: A

27. Which of the following control the menstrual and ovarian cycles of the female?
    A) GnRH
    B) FSH
    C) LH
    D) Estrogen
    E) All of the above.
    Answer: E

28. The initial secretion of estrogens by the growing ovarian follicles is directly stimulated by
A) GnRH.
B) relaxin.
C) progesterone.
D) FSH.
E) LH.
Answer: D

29. The hormone that works with estrogen to prepare the endometrium for implantation of a fertilized egg is
A) LH.
B) FSH.
C) prolactin.
D) progesterone.
E) relaxin.
Answer: D

30. The average menstrual cycle is
A) 14 days.
B) 18 days.
C) 24 days.
D) 28 days.
E) 32 days.
Answer: D

31. The rupture of the mature follicle during ovulation is due to a surge in
A) FSH.
B) LH.
C) progesterone.
D) estrogen.
E) inhibin.
Answer: B

32. The presence of which hormone in the maternal blood or urine indicates pregnancy?
A) hCG
B) FSH
C) LH
D) GnRH
E) progesterone
Answer: A

33. Currently the most prevalent sexually transmitted disease in the U.S. is
A) AIDS.
B) gonorrhea.
C) chlamydia.
D) genital herpes.
E) syphilis.
Answer: C

34. The structure between the uterus and the vagina is the
A) uterine tube.
B) isthmus.
C) cervix.
D) vulva.
E) hymen.
Answer: C

35. "Crossing-over" occurs during
A) spermatogenesis.
B) oogenesis.
C) meiosis I.
D) meiosis II.
E) mitosis.
Answer: C

36. The circular pigmented area of skin surrounding the nipple of the mammary glands is called
A) alveolus.
B) areola.
C) perineum.
D) apex.
E) vulva.
Answer: B

37. The release of estrogen and inhibin by the dominant follicle decreases the release of
A) Progesterone.
B) GnRh.
C) LH.
D) FSH.
E) Relaxin.
Answer: D

38. The absence of menstruation is referred to as
A) hypospadias.
B) PID.
C) amenorrhea.
D) dysmenorrhea.
E) smegma.
Answer: C

39. Testosterone is produced by the
A) corpus luteum
B) interstitial fluid of Langerhans.
C) hypothalamus.
D) interstitial cells of Leydig.
E) anterior pituitary.
Answer: D

39. Menstruation (the sloughing off of the endometrium) is directly caused by a decrease in the production of
A) progesterone.
B) testosterone.
C) FSH.
D) estrogen.
E) LH.
Answer: A

41. The expanded part of the uterine tube that has the fimbriae is called the
A) corpus albicans.
B infundibulum.
C) corpus cavernosa
D) corpus sponginosum.
E) perineum.
Answer: B

42. Which of the following is not true?
    A) Mitosis produces cells with identical genetic information.
    B) Meiosis produces cells with identical genetic information.
    C) Mitosis produces 2 cells with 46 chromosomes.
    D) Meiosis produces cells with 23 chromosomes.
    E) Meiosis produces 4 cells.
    Answer: B

43. Puberty is triggered by the secretion of
    A) testosterone or estrogen from the gonads.
    B) FSH from the anterior pituitary.
    C) LH from the anterior pituitary.
    D) GnRH from the hypothalamus.
    E) oxytocin from the posterior pituitary.
    Answer: D

44. Which of the following is not a form of contraception?
    A) vascectomy
    B) tubal ligation.
    C) oral contraceptives
    D) IUD
    E) All are forms of contraception.
    Answer: E

45. RU486 works by blocking the action of
    A) estrogen.
    B) HCG.
    C) progesterone.
    D) FSH.
    E) LH
    Answer: C

True-False

Write T if the statement is true and F if the statement is false.

1. The production and survival of sperm require lower than normal body temperature.
   Answer: True

2. Sperm cells mature in the testis.
   Answer: False

3. Gametes differ from all other body cells because they contain the diploid chromosome number.
   Answer: False

4. Sperm are produced at the rate of about 300 million per day.
   Answer: True

5. FSH stimulates testosterone production.
   Answer: False

6. Sperm are moved from the seminiferous tubules directly to the epididymis.
   Answer: False

7. Semen has a slightly alkaline pH.
   Answer: True

8. During early fetal development, germ cells in the ovaries differentiate into oogonia.
   Answer: True

9. The uterus is supported and held in position by several skeletal muscles.
   Answer: False

10. The mucosa of the vagina has an acid environment.
    Answer: True

11. The region between the labia minora is called the hymen.
    Answer: False

12. The secretion of milk from the mammary glands following delivery is largely due to the hormone prolactin.
    Answer: True

13. LH from the posterior pituitary causes the production of the hormone estrogen by the follicles in the ovaries but causes sperm production in the testes.
    Answer: False

14. Human gametes have half the number of chromosomes the other cells have, to accomplish this, these haploid cells are produced by a process called meiosis.
    Answer: True

15. An increase of estrogen in the blood causes and increase in luteinizing hormone which causes an increase in progesterone. Both estrogen and progesterone cause the endometrium to retain fluid, but estrogen causes the myometrium to contract and progesterone causes it to relax.
    Answer: True

16. Both start with germinal epithelium, but oogenesis is different from spermatogenesis. Oogenesis begins before birth and one egg is ovulated at a time after puberty, but testosterone begins at puberty and millions of sperm are released at one time.
    Answer: True

17. Sperm are small because they have a long way to travel. They start in the seminiferous tubules, mature in the epididymis, go through the ductus deferens to the ejaculatory duct and finally the urethra. Along the way, secretions are added by the seminal vesicles, prostate gland and bulborurethral glands.
    Answer: True

Short Answer

Write the word or phrase that best completes each statement or answers the question.

1. The branch of medicine that deals with the diagnosis and treatment of diseases of the female reproductive system is _____.
   Answer: gynecology

2. The pouch that supports the testes is the ____.
   Answer: scrotum

3. The lobules of the testis contain ______ _______.
   Answer: seminiferous tubules

4. The portion of a sperm that contains an enzyme necessary for the penetration of the secondary oocyte is the ____.
Answer: acrosome

5. The period of time when secondary sex characteristics begin to develop is referred to as ____.
Answer: puberty

6. A supportive structure of the male reproductive system that carries blood vessels, autonomic nerves, and lymphatic vessels is the ____.
Answer: spermatic cord

7. A method of sterilization in males where a portion of each ductus deferens is removed is called
Answer: vasectomy

8. The two dorsal masses of erectile tissue of the penis are the ____.
Answer: corpora cavernosa penis

9. The mass of erectile tissue surrounding the male urethra is the ____.
Answer: corpus spongiosum penis

10. The distal end of the corpus spongiosum is a slightly enlarged region called the ____.
Answer: glans penis

11. A surgical procedure in which part or all of the prepuce is removed is called ____.
Answer: circumcision

12. The erection of the penis is due to a _________ reflex.
Answer: parasympathetic

13. The remnants of an ovulated mature follicle form the ____.
Answer: corpus luteum

14. The degenerated corpus luteum is called ____.
Answer: corpus albicans

15. A growing follicle forms a clear layer of glycoprotein called the ____.
Answer: zona pellucida

16. The broad ligament that forms part of the outer layer of the uterus is the ____.
Answer: perimetrium

17. The middle muscular layer of the uterus is called ____.
Answer: myometrium

18. The innermost part of the uterine wall is the ____.
Answer: endometrium

19. The structure that serves as a passageway for menstrual flow and childbirth is the ____.
Answer: vagina

20. The region between the vagina and the anus is known as the ____.
Answer: clinical perineum

21. The mammary glands are modified ____ glands.
Answer: sudoriferous (sweat)

22. The circular pigmented area of skin surrounding the nipple is called the ____.
    Answer: areola

23. The external genitalia of the female are collectively called ______.
    Answer: vulva

24. The hormone released by the hypothalamus that effects both the female and male reproductive system is _____.
    Answer: GnRh

25. The human sexual response has ___ stages.
    Answer: 4

Matching

Choose the item from Column 2 that best matches each item in Column 1.

1. Column 1: Testes do not descend.
   Column 2: cryptorchidsm

2. Column 1: The onset of menstruation
   Column 2: menarche

3. Column 1: The absence of menstruation.
   Column 2: amenorrhea

4. Column 1: painful menstruation due to forceful contractions of the uterus.
   Column 2: dysmenorrhea

5. Column 1: Presence of both male and female sex organs in one individual.
   Column 2: hermaphroditism

6. Column 1: A displaced urethral opening.
   Column 2: hypospadias

Essay

Write your answer in the space provided or on a separate sheet of paper.

1. Describe the path of sperm through the male reproductive system.
   Answer: Sperm is produced in the seminiferous tubules of the testis, moved to the straight tubules to the rete testis. The sperm are then transported into the epididymis to the ductus deferens, where the sperm can be stored. During ejaculation, sperm is moved into the ejaculatory duct and finally into the urethra, and the terminal duct of the system.

2. Name and explain the function of the accessory sex glands of the male reproductive system.
   Answer: The accessory sex glands secrete most of the liquid portion of the semen. The paired seminal vesicles secrete an alkaline, viscous fluid that contains fructose, prostaglandins, and clotting proteins. The alkalinity of the fluid helps neutralize the acid in the female reproductive tract.
   The prostate gland secretes a milky, slightly acidic fluid that contains nutrients and several enzymes.

The paired bulbourethral glands secrete an alkaline substance that protects sperm by neutralizing the acid environment of the urethra, and a mucus that lubricates the end of the penis during intercourse.

3. Define the female reproductive cycle.
   Answer: The female reproductive cycle refers to the menstrual and ovarian cycle. The menstrual cycle is the series of monthly changes in the non-pregnant female, the ovarian cycle describes the monthly changes occurring in the ovaries.

4. Name the sexually transmitted diseases caused by viruses.
   Answer: AIDS, genital herpes, genital warts

CHAPTER 24 Development and Inheritance

Multiple-Choice

Choose the one alternative that best completes the statement or answers the question.

1. Of the sperm cells introduced into the vagina, how many actually reach the secondary oocyte?
   A) 12 percent
   B) 8 percent
   C) 5 percent
   D) 2 percent
   E) less than 1 percent
   Answer: E

2. A secondary oocyte is viable for approximately _____ hours after ovulation.
   A) 6
   B) 12
   C) 24
   D) 36
   E) 48
   Answer: C

3. Embryology refers to the study of development from the fertilized egg through the ____ week.
   A) second
   B) fifth
   C) eighth
   D) ninth
   E) tenth
   Answer: C

4. The term neonatal period defines the first ____ days after birth.
   A) 40
   B) 28
   C) 36
   D) 12
   E) 8
   Answer: B

5. Birth in general occurs ____ weeks after fertilization.
   A) 30
   B) 32
   C) 34
   D) 36
   E) 38
   Answer: E

6. Before a sperm cell can penetrate an ovum it has to undergo a series of changes called _____.
   A) capacitation.
   B) lubrication.
   C) pulpation.
   D) dislocation.
   E) maturation.
   Answer: A

7. For fertilization to occur the sperm has to penetrate layers surrounding the oocyte. The first layer is the
   A) zona glomerulosa.
   B) zona pellucida.
   C) corona radiata.
   D) corona granulosa.
   E) reticular layer.
   Answer: C

8. Fraternal twins
   A) develop from a single fertilized ovum.
   B) are the same age but genetically different.
   C) are genetically identical.
   D) are always the same sex.
   E) are fertilized by one sperm cell.
   Answer: B

9. The rapid mitotic cell division of the zygote that occurs immediately after fertilization is called
   A) development.
   B) differentiation.
   C) gravidation.
   D) cleavage.
   E) cytokinesis.
   Answer: D

10. Successive cleavages produce a solid mass of tiny cells called a(n)
    A) embryo.
    B) blastocyst.
    C) morula.
    D) zygote.
    E) trophoblast.
    Answer: C

11. The placenta is formed from the
    A) trophoblast.
    B) inner cell mass.
    C) trophoblast and inner cell mass.
    D) embryo.
    E) morula.
    Answer: A

12. The process of implantation occurs approximately ____ days after fertilization.
    A) 6
    B) 8
    C) 10
    D) 12
    E) 14
    Answer: A

13. During vitro fertilization,
    A) a woman is artificially inseminated with the husband's semen.
    B) the husband's semen is used to artificially inseminate a secondary oocyte donor.
    C) fertilization happens in a laboratory dish.
    D) a woman is given hormone treatment and is then artificially inseminated with donated semen.
    E) All of the above.
    Answer: C

14. The term fetal period refers to the time from the ___ month until birth.
    A) 1st
    B) 2nd
    C) 3rd
    D) 4th
    E) 5th
    Answer: B

15. The placenta is formed by the end of the
    A) first month.
    B) second month.
    C) third month.
    D) fourth month.
    E) immediately after implantation.
    Answer: C

16. Which of the following embryonic membranes becomes the principal part of the embryonic part of the placenta?
    A) endoderm
    B) chorion
    C) mesenchyme
    D) allantois
    E) amnion
    Answer: B

17. Which of the following serves as an early site of blood formation?
    A) yolk sac
    B) allantois
    C) chorion
    D) amnion
    E) amniotic fluid
    Answer: B

18. Wastes and deoxygenated blood leave the fetus through the
    A) umbilical veins.
    B) intervillous spaces.
    C) umbilical arteries.
    D) chorionic villi.
    E) ductus venous.
    Answer: C

19. The first fetal vessel that carries fully oxygenated blood is the
    A) umbilical vein.
    B) umbilical artery.
    C) ductus arteriosus.
    D) pulmonary trunk.
    E) common iliac arteries.
    Answer: A

20. How long is the corpus luteum maintained after fertilization?
    A) two weeks
    B) six weeks
    C) at least two months
    D) at least three to four months
    E) no longer than six months
    Answer: D

21. Which of the following develop from the mesoderm?
    A) nervous system
    B) cardiac muscle
    C) epidermis
    D) All of the above.
    E) None of the above.
    Answer: B

22. All of the following are pregnancy induced changes EXCEPT
    A) increased cardiac output.
    B) decreased gastrointestinal tract motility.
    C) increased tidal volume.
    D) decrease in respiratory reserve volumes.
    E) All of the above are pregnancy induced changes.
    Answer: E

23. Pharmaceutical companies use human placentas as a source for
    A) hormones.
    B) drugs.
    C) blood.
    D) None of the above.
    E) All of the above.
    Answer: E

24. The heartbeat can first be detected in the ____ month of pregnancy.
    A) 1st
    B) 2nd
    C) 3rd
    D) 4th
    E) 5th
    Answer: C

25. The hormone that stimulates uterine contractions is
    A) vasopressin.
    B) oxytocin.
    C) prolactin.
    D) relaxin.
    E) estrogen.
    Answer: B

26. The major hormone promoting lactation is
    A) oxytocin.
    B) vasopressin.
    C) prolactin.
    D) estrogen.
    E) progesterone.
    Answer: C

27. Milk ejection is promoted by
    A) oxytocin.
    B) vasopressin.
    C) prolactin.
    D) estrogen.
    E) progesterone.
    Answer: A

28. Genes that control the same inherited trait that occupy the same position on homologous chromosomes are called
    A) carries.
    B) alleles.
    C) dominant.
    D) recessive.
    E) equal.
    Answer: B

29. Which of the following is caused by a dominant gene?
    A) sickle cell anemia
    B) fetal alcohol syndrome
    C) Down's syndrome
    D) Huntington's disease
    E) phenylketonuria
    Answer: D

30. The nuclei of all human cells, except gametes, contain
    A) 46 pairs of chromosomes.
    B) 36 pairs of chromosomes.
    C) 28 pairs of chromosomes.
    D) 23 pairs of chromosomes.
    E) 22 pairs of chromosomes.
    Answer: D

31. The actual genetic makeup is contained in the
    A) phenotype.
    B) genotype.
    C) Punnett square.
    D) dominant genes.
    E) All of the above.
    Answer: B

32. Which of the following is a teratogen?
    A) cocaine
    B) alcohol
    C) nicotine
    D) pesticides
    E) All of the above.
    Answer: E

33. Which of these diseases is a recessive condition?
    A) cystic fibrosis
    B) diabetes insipidus
    C) Huntington's disease
    D) Both A and B.
    E) All of the above.
    Answer: A

34. Which of the following is a condition that is inherited by sex-linked traits?
    A) Huntington's disease
    B) diabetes insipidus
    C) cystic fibrosis
    D) hemophilia
    E) PKU
    Answer: D

35. Klinefelter's syndrome is a sex chromosome disorder with the following genotype:
   A) XYY
   B) XXY
   C) XXX
   D) XO
   E) XY
   Answer: B

36. After the morula has formed a fluid formed cavity it is called the
   A) morula II
   B) chorion.
   C) blastocyst.
   D) fetus.
   E) bilaminar embryonic disc.
   Answer: C

37. Fertilization is most likely to occur
   A) within one week of ovulation.
   B) between 2 days before and 1 day after ovulation.
   C) one week before ovulation.
   D) 4 days prior to and 2 days after ovulation.
   E) just before implantation.
   Answer: B

38. Which layer secretes enzymes to help the blastocyst penetrate the uterine lining?
   A) syncitiotrophoblast layer.
   B) cytotrophoblast layer.
   C) endoderm layer.
   D) mesoderm layer.
   E) chorionic layer.
   Answer: A

39. The primitive ectoderm is also known as the
   A) mesoderm.
   B) endoderm.
   C) notochord.
   D) epiblast.
   E) hypoblast.
   Answer: D

40. A chart which uses the possible paternal and maternal gene combinations to show the probability of all genotypes in the offspring is called a(n)
   A) phenotype.
   B) allele table.
   C) Punnett Square.
   D) dominant trait.
   E) incomplete dominance.
   Answer: C

True-False

Write T if the statement is true and F if the statement is false.

1. Monozygotic twins develop from a single fertilized ovum.
   Answer: True

2. The zona pellucida is a layer of clear glycoprotein around the ovulated oocyte.
   Answer: True

3. The months of development after the second month are considered the fetal period.
Answer: True

4. The allantois becomes the principal embryonic part of the placenta.
Answer: False

5. The placenta and the corpus luteum produce the hormone relaxin.
Answer: True

6. The principal stimulus in maintaining prolactin secretion during lactation is the sucking action of the infant.
Answer: True

7. Following birth, prolactin levels start increasing.
Answer: False

8. Homologous chromosomes contain similar traits.
Answer: True

9. The sex chromosomes are also responsible for the transmission of a number of nonsexual traits.
Answer: True

10. Maternal smoking has no impact on the developing fetus.
Answer: False

11. X and Y chromosomes are sex chromosomes.
Answer: True

12. PKU is a sex linked inherited disorder.
Answer: False

13. A sperm has a tough job if it wants to fertilize an oocyte. It must survive the acid vagina, be capacitated, select the correct uterine tube, penetrate the corona radiata and be the first sperm to do so.
Answer: True

14. If separation of monozygotic twins does not occur after about a week, conjoined twins are likely to occur.
Answer: True

15. The primitive spinal cord develops from the neural tube.
Answer: True

16. The yolk sac, which develops from the blastocyst cavity, has the functions of providing nutrients to the early embryo and supplying blood cells in the second three weeks of development.
Answer: True

17. Even though most organisms cannot cross the placenta, several drugs and viruses can cross that may cause birth defects, especially in the embryonic stages.
Answer: True

18. True labor can be divided into four stages, the stage of no show, stage of dilation, stage of expulsion, and the placental stage.
Answer: False

Short Answer

Write the word or phrase that best completes each statement or answers the question.

1. Twins that develop from a single fertilized ovum are called ____.
   Answer: identical or monozygous

2. When the morula forms a hollow ball of cells, it is then called ____.
   Answer: blastocyst

3. The outer layer of a blastocyst is called the ____.
   Answer: trophoblast

4. If an embryo implants in a location outside the uterine cavity it is called an ______ pregnancy.
   Answer: ectopic

5. The three primary germ layers are the ectoderm, endoderm, and ______.
   Answer: mesoderm

6. All nervous tissue develops from the _______.
   Answer: ectoderm

7. A procedure in which a husband's semen is used to artificially inseminate a fertile secondary oocyte donor, followed by a transfer of the morula to the infertile wife is called _____.
   Answer: embryo transfer

8. The ectoderm separates to form a fluid filled space called the ____.
   Answer: amniotic cavity

9. During embryonic life, finger-like projections grow into the endometrium of the uterus. These projections are called ____.
   Answer: chorionic villi

10. After the birth of the baby, the placenta detaches from the uterus and is termed the _____.
    Answer: afterbirth

11. Blood passes from the fetus to the placenta via two ____ in the umbilical cord.
    Answer: umbilical arteries

12. The period of time a zygote, embryo, and fetus are carried in the female reproductive tract is called ___.
    Answer: gestation

13. The process by which uterine contractions expel the fetus through the vagina to the outside is called ____.
    Answer: labor

14. The term emesis gravidarum means ______ ______.
    Answer: morning sickness

15. The secretion and ejection of milk by the mammary glands is called ____.
    Answer: lactation

16. During the first few days after birth, the mammary glands secrete a fluid called ____.
    Answer: colostrum

17. The branch of biology that studies inheritance is ____.
    Answer: genetics

18. The way the genetic makeup is expressed in the body is referred to as ____.
Answer: phenotype

19. Inherited traits that are not controlled by one gene, but by the combined effect of many, are referred to as _____.
Answer: polygenic inheritance

20. Intrauterine exposure to alcohol results in ____.
Answer: fetal alcohol syndrome

21. The chromosomes, which are not sex chromosomes, are called ____.
Answer: autosomes

22. A permanent change in a gene that might be inheritable is a ____.
Answer: mutation

23. An abortion that occurs without apparent cause is a ____.
Answer: spontaneous abortion

Matching

Choose the time form Column 2 that best matches each item in Column 1.

1. Column 1: tubal ligation
Column 2: sterilization

2. Column 1: vasectomy
Column 2: sterilization

3. Column 1: oral contraception
Column 2: hormonal method

4. Column 1: a small object of plastic, copper, or stainless steel inserted into the uterine cavity.
Column 2: intrauterine device

5. Column 1: condoms
Column 2: barrier method

6. Column 1: vaginal pouch
Column 2: barrier method

7. Column 1: cervical cap
Column 2: barrier method

8. Column 1: diaphragm
Column 2: barrier method

9. Column 1: spermicidal agents
Column 2: chemical method

10. Column 1: rhythm method
Column 2: natural method

11. Column 1: sympto-thermal method
Column 2: natural method

Essay

Write your answer in the space provided or on a separate sheet of paper.

1. Name the primary germ layers and the main structures developing from each.
   Answer: 1. Endoderm: it becomes the epithelial lining of the gastrointestinal tract, respiratory tract, and a number of other organs.
   2. Mesoderm: it forms the peritoneum, muscle, bone, and other connective tissue.
   3. Ectoderm: develops into the skin and nervous system.

2. Describe the stages of labor.
   Answer: 1. The stage of dilation is the time from the onset of labor to the complete dilation of the cervix.
   2. The stage of expulsion is the time from complete cervical dilation to delivery.
   3. The placental stage is the time after delivery until the placenta is expelled.

3. Name the principal methods of birth control, and give an example of each.
   Answer: sterilization- vasectomy and tubal ligation
   Hormonal methods- oral contraceptive
   Intrauterine devices- copper T
   Barrier methods- condoms
   Chemical methods- spermicides
   Natural methods- rhythm, coitus interruptus, induced abortion